The Quiet Tree
How Australia's w
save the planet

By

Leslie H. Harvey of Kimberley

Copyright © Leslie H. Harvey of Kimberley 2021

The right of Leslie H. Harvey of Kimberley to be identified as the Author of the Work has been asserted by him in accordance with the Copyright, Designs and Patents Act 1988.

This publication may only be reproduced, stored or transmitted, in any form, or by any means, with prior permission in writing from the author.

A CIP catalogue record for this title is available from the British Library

ISBN: 9798542523934

Acknowledgments to the amazing Jessica

If you would like to contact the author please feel free to email him at
lesliehharvey@outlook.com

Part 1.
Notes on Australia.

by

Leslie H. Harvey of Kimberley.

Australia has the secret of the future for our Planet, and maybe the little town of Bingara in NSW can be the archetype of that future.

Walkabout - an informal stroll.

You will come across five important themes in this work, and I thought it prudent to give you a head start by mentioning them at the beginning to prepare you for you walkabout.

Global warming:

The last few decades of global drying do bear indications of global warming, and scientists have verified a pole ward shift in the storm track that girdles the Southern Ocean.

Myall Creek:

This was the site of the massacre of 25 to 30 Aborigines in the 19th century, just one indication of ill treatment by white folk on indigenous peoples for the sake of supremacy.

Small town mineral wealth:

In places like Bingara, it was the discovery of gold in the land of Australia in 19th century, along with copper and diamonds, which made unregulated rapid change inevitable.

A non-alliance with politicians:

It would seem to me that the politicians have only one aim in life, their self preservation, no matter what it costs.

Aborigine Life:

The aborigine way of life as been proved over 80,000 years, they made it, we stand a very 'good' chance of not making it.

Author's Extra Note: This Planet is our only home, and the only chance we have, damage it and we run the risk of losing everything, maybe our own future, but more certainly the future for our children and their children, an about turn is needed now.

27 March 2018

Have an honest walkabout around Australia with me; while I 'wheelabout', and stay where you wish, it is a big country after all.

Chapter 1. Captain Cook.

8 October 2018

01. Captain Cook 1:

In his words:

'They, the Aborigines, may appear to be the most wretched people on earth but, in reality, they are far happier than we Europeans are. They live in a tranquillity that is not disturbed by the inequality of conditions; the earth and sea of their own accord furnish them with all things necessary for life'.

02. The Endeavour:

A Captain Cook observation:

The creaking *Endeavour* was clearly the largest and most extraordinary structure that could have come before them, 'yet most of the aborigine natives merely glanced up and looked at it as if at a passing cloud, and returned to their tasks'.

03. Captain Cook 2:

Another observation.

'The aborigines may appear to some to be the most worthless people on earth, but in reality they are far better off than we Europeans. They live in a way that is not disturbed by the inequality of conditions.

The earth and the sea, of their own accord, furnish the aborigines with all things necessary for life.

All they seem to want is to be left alone'. Anon.

04 November 2018

04. Botany Bay:

Botany Bay was the destination of the first convict fleet. However, on arrival there, the bay was found to be unsuitable and the settlement was made instead at Sydney Cove.

Chapter 2. History.

05. The Founding of White Australia:
On the 26th of January 1788, Captain Arthur Phillip R.N. planted a flag on the beach at Sydney, and this is now Australia Day; and don't you forget it. (Invasion day to some)

7 October 2018.
06. Discovery:
The earliest European visitors, probably Dutch, who landed on the west coast of Australia in the 17th century, never explored further than the coastal dunes and were unenthusiastic about the region's potential. It wasn't until 1829 that the first British settlers arrived in the Swan River Colony, later Perth.

07. Settlers 1. They did their best:
On the edge of a barren wasteland far to the north of Melbourne, where Settlers did their best to settle, and around a thousand kilometres from the sea, there is an abandoned well with nothing to show of why it is there.
However, it does serve to commemorate that first family that tried to settle there, and if you should come across such a well, pause and think on, and remember, even though the place today is more of a barren region of mistakes; abysmal efforts dealt the final blow of failure, it was once somebody's dream.

25 April 2018

08 Anzac Day:

26th of April. The day we all should always remember no matter where we are, even in Kimberley, a small town in the county of Nottinghamshire, in the middle of England, a 25-hour flight from the Balarat Anzac Memorial, and 10,511 miles from Melbourne.

Author's Extra Note: The war memorial in Balarat Australia is very similar to the war memorial in Kimberley UK

23 August 2018

09. Thomas Muir c1794:

My mind tells me that I have acted agreeably to my conscience and that I have engaged in a good, a just and a glorious cause – a cause which will save this country from destruction.

I will paraphrase that:

My mind tells me I am acting agreeably to my conscience and that I am engaging in a good, a just and glorious cause which will save this planet from destruction.

10. Regular Folk:

I won't call them common or normal folk, no, they, like me, are regular kind of people, good and kind but willing to get right up somebody's nose if need be, but even then, not with any sense of malice. Let's see how it goes shall we.

11. Deportation:
To escape the atrocious conditions in Britain, many committed petty crimes in order to be deported to Australia, where they thought conditions could be no worse, or maybe even better.

12. Wool:
Wool became paramount in the Australian economy from 1797, and stayed that way for a long time.

13. Settlement:
In 1815, the year Napoleon was crushed and Britain could once more turn her strength to building an empire, the map of white settlement in New South Wales was hardly more than a patch, consisting only of Sydney and Parramatta.
The non-white part of New South Wales was a void, the scarcely penetrated green continuum of bush, with a few tracks winding their frail, dusty capillaries inland.

Author's Extra Note: We have a primitive chance to be more sophisticated than we have ever been.

13 November 2018.

14. Enforced penal settlement:

The first penal out-station on the mainland was Newcastle, founded where the Hunter River flowed into the Pacific about seventy miles north of Sydney.

15. Early Survey:

Oxley and his son reached Moreton Bay by sea and carried out a simple survey. They found a river, rich soil, plenty of fresh water, and friendly Aborigines.

16. Free Australians:

Free Australians "We were not at a state of advancement to be benefited by separation... We possessed little of the stern and sturdy spirit of the old American colonists." Henry Parkes

17. Western Australia:

Except for some coastal patches, it was all desert, pebbles, saltbush and spinifex – the right spot in the Australian phrase, *'to do a perish'*.

18. Downturn:

Toward the end of the 19th century, sheep that had cost £4 (10A$) to £5 (14A$) a head were going begging at 2/6d (6c). The price of their wool had plummeted, leaving the grazier no margin at all, and real hardship became the norm.

22 November 2018.

19. The Beginning of Agriculture:

It was not until 1813 that the divide of the Blue Mountains was crossed, by Blaxland, Wentworth and Lawson, and then the Australian agricultural potential was realised.

20. Resettlement:

Around 1836, with the building of a railway, the people began building elegant homes and hotels, in the mountains, refuges from the stifling summer heat of the coastal places.

21. Blue Mountains:

Nowadays the Blue Mountains are just 2 hours away from the city by road and rail, and not the two or three days they once were.

22. Edward Hargraves:

In 1851, Edward Hargraves went prospecting for gold west of Sydney on land beyond the Blue Mountains, around Orange and Bathurst, and found gold in abundance, and an Australian gold rush began.

In less than ten years, the population of the State of Victoria expanded wholesale, with thousands of new faces.

23. Rabbits: (and counting)

In 1859, Thomas Austin imported twenty-four rabbits into Winchelsea Victoria, and released them into the bush for sport, and within a couple of years, the area was overrun with rabbits. By 1880, the rabbits had picked clean around two million acres of the state of Victoria, and Mr Austin had disappeared down a rabbit hole me thinks.

The rabbits then spread into the states of South Australia and New South Wales at a rate of seventy miles per year.

Before the rabbits, the countryside in these states was characterised by lush groves of emu bush which was in flower most of the year. Now there are no flowering bushes and no green swards, just Mr Austin looking for a place where there are no rabbit 'tods'. Keep looking mate.

24. Sheep:

As the rabbits moved ever onwards, the sheep lost their feeding grounds and died off in huge numbers, and when a decade of drought hit the regions, some 35 million sheep perished; in the year 1902 16 million sheep were lost. If five rabbits equal one dead sheep, how many rabbits were nibbling with a smile on their face in 1902?

25. History and Future of Australia:

What happened to the Australian hinterland in the late 19th and early 20th centuries, the overgrazing by rabbits, sheep and cattle, and the prolonged drought of that period, is what will happen to this world of ours if we do not see sense and act now. Of course, we can all go on holiday and forget all about it, until the tide comes in with a vengeance.

26. Farming:
In the early days of European farming in Australia, the future effect of what was happening was never contemplated, or if it was, it was ignored. An imprudent farmer ain't no farmer at all, then or now.

27. Settlers 2:
European settlers of the Australian variety of the last two hundred years or so were in a pinprick of time in the settlement of Australia. The 80,000 or maybe more years before that short period, people came from another part of the world to settle successfully in a huge land.
They farmed successfully and brought life to the place in abundance of self-regulating natural crops and animals.

Author's Extra Note: It is my belief that if we should return to the simple farming techniques practised by the 'before time' settlers, but retain some clever natural advancements, we would be in a good position to save Australia and this Planet. A 'plastic future' is no future at all.

28. Settlers 3:

Settlers of the Australian variety, either settlers of the 19th and 20th centuries or those of a generations spreading over thousands of years, are to be admired for their downright determination to succeed – and that is why I claim they have the right to lead us out of the predicament we find ourselves in today. They can lead the way, and help save the planet in fact.

29. History in Pictures:

Drawings made by the early observers should be viewed with an open mind. They were white 'surveyors', British explorer/settlers who deemed it necessary to belittle the lifestyle of the Aborigines, a defining process that is still in evidence today, sadly.

29 December 2018.

30. First Fleet:

To say they set sail purely for the sake of discovery is tantamount to blasphemy; they set sail for Australia to find somewhere to dump the unwanted people* the British society seemed to breed at the time, thereby making Britain cleaner at a cost that no one dared calculate, or could not be bothered to.
*Unwanted by the powder puffed high society folk, those who had about as much decency as a camel fart.

Author's Extra Note: Against all odds, the new settlers got stuck in, and the new country flourished, and is probably the cleanest nation on the planet now, but it is in danger of joining the 'old lot' by adopting detrimental practices fed by greed.

31. The Scottish Highland Clearances:

A well to do Laird declared his desire to turn his crofting land into sheep grazing, and, as a consequence, he didn't give a shuttle about his tenant crofters being made homeless.

For a time, the Highland Crofters struggled on poor land by the sea, but the circumstances were too fierce and some of the women and children died of malnutrition.

Taking life by the scruff of the neck, some of the Highlanders emigrated to Canada, and some to Australia.

Scottish settlers in Australia had to be tough, and they had to make tough decisions, and survive they did, in a settlement by default – sheep were the primary cause for their enforced emigration, and the same animal served to make amends in Australia. But then things went hacienda up.

32. Only a memory:

Time has passed and what was once real is now a memory, but every day we live is a gift from those who are fading from memory. Let not that memory be vandalised or their gift to us wasted.

They gave their loves and left a memory, each one a man or woman who was once a physical part of Australia.

Their bravery, suffering and sacrifice must not be wasted, make the memory of them grip us by the hand and shake us into a new revelation.

They may only be a memory, but let that cling in many ways, and make their actions count. They meant us to carry on in peace and good welfare, just bloody get on with it.

Chapter 3. Outback.

33. Outback Dry:

In the Outback in the dry season, when the climate is ultra comfortable, the creeks and floodplains will have dried up, concentrating wildlife on the permanent wetlands. This is a good time to see crocodiles and birds, but the land is drier than most would expect, and, please, remember you are a guest, even if you are a permanent inhabitant.

34. Outback Wet:

In the wet season, the Outback is most dynamic. Road access is restricted to a few areas only, and the humidity is uncomfortable with frequent rain, sometimes very heavy rain, but the land is in its best green state. Don't be a pillock, stay out of it, unless you know what you are doing or you have a guide.

7 December 2018.

35. Outback Travel:

'There was a massive sand hill and a forest of desert oaks in a valley, and when I prepared to leave it, I paused to see it again. I had not expected anything quite so naturally beautiful'.

Author's Extra Note: What must it be like to stand in the middle of emptiness, and know that you are probably the only human for kilometres (or miles) and days? Of course, you are closer to the land beneath your feet than you are to the sky, and you are closer to the earth-spirits than you are to another human.

36. Inland Australia:

There was nothing but sandhills, spinifex, and interminable space. I was perhaps treading now on land where no one had ever walked before, there was so much room - pure virgin desert, not even the odd cow or two to mar it, and nowhere in that vastness, was there even an atom of anything human. Keep looking mate.

37. In the words of Banjo Patterson:

He camped in a Billabong, under the shade of a coolibah tree, and sang as he sat there, looking at his old billycan boiling, thinking, who'll come a waltzing with me. He was on his own, out in the bush, and he was experiencing the calm of being there, his only companions a billycan, and a coolibah tree. Dedicated to Banjo, in all probability, he knew the real Australia, and like many others, he loved it.

38. 'Bush' Poet:

A bush poet living in the heart of England, and in the middle of winter, is about as barmy as it gets, but then, I could be waltzing with Matilda to keep warm.

39. In the Quiet:

Out in the relative quiet of the open country, far from the hidden or disguised noise of traffic and socialising of the towns and cities, time still moves on as it did thousands of years ago, or even just a couple of hundred years ago, but it does so politely, and smoothly, and mostly in a quiet manner.

However, it is never silent; the 'quiet' can be codified in many varied ways. Drop a stone onto hard ground, and the noise will be sharp if only momentary. However, the wind, the movement of air around a protruding object, a bush or tree, or a rock, will make a seemingly everlasting sound, but the open world where you are will still remain quiet, but for how long.

40. Bush Food:

'Bush food is for free, ask an Aborigine, and blooming well take heed if you want to survive, and that's not just in the bush.' A comment made by an old bushranger who had been 'close' more than once.

41. Plains:

An extensive area of level or rolling treeless country, to make a broad expanse of unbroken land.

42. Foothills:
As the name implies these are hills or a range of hills at the bottom of a mountainous country. The low country between the mountains and open country or valley, where the land changes to match the surroundings. But they are something a lot different to an Aborigine.

43. Space:
Space. loads of it, seemingly limitless and virginal in its entity, but the Aboriginal world is all within walking distance, whichever way you put it, or where you put your next stride, and for how long you can walk. In Aboriginal terms, it's not miles or kilometres, it's just distance.

44. Washaways:
 Not a waterway, but an area that can be carpeted in green and dotted with white, yellow, red, blue wildflowers. A creek-bed where tall gum trees and delicate acacias can cast deep cool shade' for free.

45. Taming the Land:
The Aborigine way is an easy and calm set of practical applications, a concept that many of us not of the Aborigine find difficult to comprehend.
This is a woeful and deadly phenomenon, and one that has lasted since 'white' feet first set foot on the Australian shore.

We of the modern social world have failed to learn and adopt the basic Aborigine truth, that the land is not ours, it is we who belong to the land - get that right and we will be somewhere on to knowing the truth of the planet we live on. Get that truth right and we will be on the way to saving our planet, our home and the home of those yet to come. More gobbledegook, or commonsense? Just leave things as they are and let our descendants lament the fact that we did nothing.

Author's Extra Note: If you put a dollar on a shovel full of good soil in the middle of Australia, which is the most valuable, the money, the shovel or the soil?

46. Movements and Patterns:

The movements and patterns in the Australian bush are of a multiplex 'multiplicious' mix we are only beginning to understand, to know and experience. A motion or pattern may only be a speck of realisation at the very limit of our vision, either aesthetically so, or on the distant rim of actual definition. Even so, we should check, and check again until we get it right. We can then begin to learn to cherish what we find, and take the next step to a sweet understanding that makes sense.

47. Mungilli Claypan:

There is shade everywhere and the sand is a soft summer rose colour. There are giant ghost gum trees are shining, and the birds are singing their own sweet tunes or chirping a conversation.

All around is emptiness, a flat expanse of hot emptiness, except for the odd tree, there is nothing of any significance to the untrained eye.

But there is life; the smooth trunks of the trees reflect the morning and evening light in a weird reassuring way, while the leaves shimmer a tentative friendly greeting.

48. Warburton:

Warburton is on its own, if there were any trees, they have long gone to firewood.

If there was any grass the cattle have eaten it and boodled off, leaving a nice covering of dust everywhere.' Of Course, there are the flies, bloody billions of 'em.

49. Money:

To have money you must first have a reason to have money, take away that reason and money becomes of no use.

What good is money in the outback, out in the bush, when all you need do is walk a bit, look around, sit down, and pick up all you require for sustenance as you go.

If what you need is freshly available to you, and free, money is without any value. Many years ago, money was not in existence in Australia, not required and not available, and when the end is the end, it will be bloody useless anyway.

Chapter 4. Bingara.

01 November 2018

50. Bingara: (My adopted town in Australia)

Bingara is 971ft above sea level; from what I can see from photographs, it would seem that the town is near to mountains and hills to the east and plains and bush lands to the west.

I know there are mountains and hills to the north and east, and open country to the west and south, which means there is a possibility of forming a lake with a water supply from the Gwydir River.

Why a lake? It's much better than a mini desert I would have thought.

15 November 2018

51. 'Small Town Australia'. Bingara, revisited:

Bingara is a small town on the Gwydir River in Murchison County, in the New England region of New South Wales, Australia.

51 (a). Bingara is currently the administration centre for the Gwydir Shire that was created in 2003.

The Gwydir River is the main highlight in the town and is a main catchment of the Murray-Darling System.

51 (b). Bingara is located 141 km north of Tamworth, 54 km west of Inverell. 449 km north of Sydney and 358 km south west of Brisbane, and very close to Myall Creek, the site of the massacre of 25 to 30 Aborigines

52. Bingara Diamonds:

Bingara is one of the places in Australia where diamonds have been found. In fact, Bingara was the largest producer of diamonds in Australia at the time.

Bingara changed the spelling of its name from *Bingera* to Bingara in 1890.

The first *Bingara* Post office opened on the 1st of January 1853 and was renamed Upper *Bingera* Post Office in 1862 and closed in 1868. The second *Bingera* office was opened in 1862 and was renamed Bingara in 1890.

53. Bingara Tourist Centre:

Tourist Officer: Jenny Head

Roxy Theatre

Maitland Street

Bingara

NSW 2404

Australia

54. Bingara Advocate:

Editor: Nancy Capel

34 Maitland Street Bingara

PO Box 15

Bingara

NSW 2404

Australia

https://catalogue.nia.gov.au

55. Notes on Bingara:

Established 1840

Population as of 2018 1,428.

Postcode 2404

Elevation 296 metres above sea level

LGAs: Gwydir Shire

County: Murchison

Federal: Parkes

Coordinates: 29'52.02' South 150'34.0' East.

56. Bingara Heritage Listings:

74 Maitland Street: Roxy Theatre and Peters Creek Cafe Complex.

https:// www.bingara.com.au

57. Bingara Population:
According to the 2016 census of Population, there were 1,428 people in Bingara. 83.7% of people were born in Australia and 88.2% of people only spoke English at home
Bingara is a popular site for retirement with 57% of the population aged 55 years and over, compared to the national average of 27.6%, and a median of 61.
The median weekly income per household is $743 (£437) which is lower than the national median of $1,438 (£845)

58. Bingara Religion:
The most common response for religion were Anglican 38.6%, Catholic 16.8%. No religion 15.2%.

59. Bingara Climate:
Bingara has a climate of spring in September and October and autumn in April and May.
During the winter months of June to August the days are sunny while the nights are cold and frosty. Summer days during November to January are dry and hot with low humidity.

60. c2314 The Bingara Renaissance:
Centuries had passed and many would be planet savers had long been dead, and the folk left alive were struggling to come to terms with the fact that the planet was on the very brink of its own death. However, just to the west of Bingara, out in the

desert regions, a new township and sprung up, New Bingara, and the folk of that town had found the answer to the global problem.

61. c2315 The Bingara Retort:

Keeping all matters to a natural level of commonsense, the folk of New Bingara had constructed a water retention scheme that turned the river on its head, and made it flow inland and not out to the sea.

Within ten months, they had found and made good a new inland lake of freshwater, Lake Bingara, a accumulation of freshwater of over a thousand square kilometres. This lake was now attracting people and animals, and the land thereabouts was beginning to return to its pre 18th century condition.

Author's Extra Note: After much investigation, a new additional water-retaining scheme was in force, and the water of Lake Bingara was spreading further into the desert, with trees planted in their thousands.

62. Myall Creek 1:

Well, you have to understand, to look at it, there is nothing all that special about Myall Creek, until you remember Aborigines who were roped together there and slaughtered. Remember them.

62. (a)

Three months before the Myall massacre, 200 Aborigines were killed at Waterloo creek, near to Moree'. Remember them. Moree is sixty or so miles west of Myall, and although 200 Aborigines perished, nobody was ever punished - the law of the day did not even *try* to punish anyone for the 'crime'.

63. Myall Creek 2:

'There's never been an archaeological investigation, and there's no memorial of any kind,' and no graves. But they do have this old man's reverential mention. Remember them.

By the way:

64. Australian Camels:

Towards Emu Field, across 150 kilometres of peak-and-trough sandhills, camels could be spotted milling in mobs between the trees. Camels and trees in unison: like the Aborigines, the camels know how to survive.

Chapter 5. Settlers and Aborigines.

02 January 2019

65. The Mix of 'Settlers' and Aborigines:
When the Settlers landed in Australia 200 or so years ago, life for the Aborigines changed. It was a dramatic change, and, in many thousands of cases, with fatal results.

 The original natives of Australia had been on the land for maybe 80,000 years, living in isolation as near to perfection as humankind can get. They could not handle the 'invaders' and their derogatory way of life, it was very alien to them, and in consequence neither could the invaders understand the native's imposing of a relaxed and near idyllic existence.

The resulting clash was, and still is, difficult to accept, and more time is maybe needed for a full integration to take place. However, the one criteria that might impose a sense of urgency is the onset of global warming, a definite actual threat to life, with only the Aborigines having the real answer. If only the so-called intelligentsia could grasp the reality of life at a basic 'intelligent' socially acceptable level, humankind might stand a chance of overcoming the deadly problem that time is running out, fast.

66. Screen Australia:
See the places of interest without getting of your beam end. Sit in your armchair and dabble in a bit of Australian socialising on the web. It might not be legit, but at least it's better than nowt (nothing).

67. Physical Links:
A constant reminder of who we came from, our ancestors, the knowledge of them making a strong physical link with the past. We have the same hands and feet, the same eyes and ears, and we are able to make out what is right and wrong; so why do we faff about with the wrong attitude and the wrong kind of actions? The enigma of human frailty perhaps.

68. The medium of Television:
The news of global warming breaks; a real threat to our planet, and for a time the TV spits out well meaning pictures and dialogue. Then another news item draws the attention; maybe some politician as trod in his or her own shuttle, and the topic of global warming fades away.
What the TV news people dedicate to the screen is becoming more of an opinion than actual news, and we are being verbally shoved up a garden path that only leads us to more unbridled ignorance, then there will be a new type of news, and opinions will change.

Switch off the TV and look out the window, you will get a more honest sense of news that way, or, better still, open the door and go and have a look for yourself.

69. The Profit Motive 1:

Good farming land has the potential to be profitable, and according to the early settlers, for Aborigines to have the land was a waste of potential, so the profit motive ruled supreme, as it does so to this day.

The profit motive works for the few, or even just the one, whereas the non-profit motive sees a modicum of fairness that benefits the majority. This is not just sense, it's *good* commonsense.

70. In the Dreamtime Zone:

Everything in the universe, and in the desert I would suppose, is in constant interaction with everything else, all rubbing together nicely, until the boffin 'wallys' get going that is. Show a boffin wally's a problem, and he or she will definitely find a way to upset it. They will then proclaim to the rest of us how clever they are by showing where *'we've'* gone wrong.

71. The Profit Motive 2:

Good farming land has the potential to be profitable, and in 18th century European eyes, for Aborigines to have the land would be a waste of potential. That was the driving force behind most of the first settlers to come from Europe, their own hapless degradation of an upbringing made it imperative that they 'make it' in the new world of Australia.

However, and sadly so, this essential force of retribution-fed greed made life impossible for the original settlers and the Aborigines, and the latter were forcibly removed from their homes and

homeland, for the profit motive become the word of law. Then the use of a verbal dress up of words was often used, even though on many occasions it was outside the law.

72. Sitting under a tree:
With your back to a tree, you are exercising an old safety measure; your enemy can only approach you from the front. However, that may depend on the size of the tree, but maybe that's being a wee bit pedantic. But if I am safe why should I give a shuttle about being finicky?

73. Automatic Watering:
Automatic watering in the natural situation is normally called rain, a phenomenon that is not readily available in most areas of the great expanse that is inland Australia.
However, there is a way around this lack of life promoting rain, but it will need a complete change in the thought processes dealing with this.

This may be considered a rambling consideration of no real value, but it should be remembered that the planet is under a real threat of extinction, and any salvation offered should be welcomed, even if that offer is from a source hitherto considered unintelligent, first settlers of Australia, the Aborigines.

74. A shining warmth:
It is Sunday May 7th 2017, and 9.15 in the morning in Kimberley, England. The sun is shining and creating a warmth that we can all enjoy, but, in parts of Australia, the sun is not so friendly, it can be a killer, the purveyor of a slow death, a hot drying killer of that part of the planet
The only defence against this is water, water we have in plenty in England; all we need do is move it from one place to another, from one area of the planet to where it is sorely needed to sustain life.

Author's Extra Note: If political 'actions' are involved in this envisaged movement of water, more's the bliidy pity.

75. Secrets:
That which is secret is of more importance than that which is known to all.
Even as I write, some secret endeavour is underway, or to be more accurate, a number of secrets are being formulated all

over the planet. I do not say I am privy to these secrets, I am just aware of human society and its foibles.

76. Masonic:

Masonic dignity is based on secrecy, and it is this that makes change difficult. In the open, a secret loses its desirability, its commercial worth is diminished. However, the fewer that know, the more commercial acumen can be applied for the benefit of the few. This is non-existent in the Aborigine world, their dignity is founded on something more natural.

77. Science v Mother Nature:

Each baffling predicament of a scientific nature is treated with qualified remedies. That's all well and good on the face of it. However, for every one scientifically base reaction, there are numerous more naturally based remedies that could easily outstrip those projected by the scientific/business folk, but this weighty science/business marriage promotion makes inroads into commonsense and the natural remedy is pushed away.

78. Embarrassment:

Being embarrassed at another's folly or even stupidity is a reaction accepted for its honesty.
However, when that embarrassment hides the real significance of an ability to recognise commonsense, this is ignored by a modern society's acceptance of the planet's certain death. It is

as if they are too embarrassed to admit that they might be wrong.

79. Grammatical and political correctness:

Being grammatically and politically correct is about as much use as trying to shift water with a fork when a flood is approaching. Grammar and politics do mix, but the correct vowel or consonant may not sway the argument in what needs to be said if it's too blooming late, and the planet is dying.

Out an about:

80. Adelaide 1:

This is a leafy city on the River Torren, with a laid-back lifestyle, and set against the hills of the Mount Lofty Ranges, and is complete with fine colonial buildings.
There is a State Library in Adelaide with a good collection of Aboriginal paintings and other artefacts.

81. North Adelaide Heritage Group:

www.adelaideheritage.com.au
There are 21 self-contained historic properties within distance of Adelaide Business District.
Accommodation ranges from a cosy 19th-century cottage to a former fire station, complete with a 1942 fire engine.

All properties are beautifully furnished and breakfast provisions are included in the price.

82. Adelaide 2:

A city with parks of green is to be envied by the smaller towns in the arid areas further inland.

All there is between Adelaide and the South Pole is a blood cold sea and a huge amount of ice, with a penguin or two and definitely no rabbits.

Chapter 6. Environment:

83. New Farms:
Ground hardened by animal and human ingress will not hold any rainwater that might fall. Because of this, the river will rise and the rainwater will wash along the riverbed, and eventually run out to sea. This outgoing of freshwater is undeniably a heartbreaking waste of a source of nutrients that would have hitherto made the new farms and farmland a much more viable proposition.

84. New Tree Plantations:
You don't have to be an environmentalist to recognise the true value of a forest of trees in the land towards the centre of Australia. Neither does it take much for those who think it a load of nonsense to suggest such a thing, just an effort to wind up their ignorant holier than thou way of thinking. They will then cringe all the way through the dry inland areas, rather than admit they were wrong

85. Trees come from sand:
The seed of a tree starts to grow when it is wet, even in sand. over long periods of time trees van change, when the DNA of a small tree changes it may well promote a larger and stronger variety of the same tree.

These larger and stronger tress then produce seeds that contain the right DNA to ensure a continual spread of the larger and stronger trees.

86. Different kinds of plants:
Many different kinds of plants of millions of years ago became trees, and these made the first forests of immense sizes.
The Australian trees have evolved in a rare and somewhat odd environment, and they would not survive elsewhere.
We need to take care of all trees; it is not a stretch of imagination to say that without trees we would not survive.

87. Commonsense from anywhere:
Does it matter where it comes from, or who it is that proclaims it, if it is commonsense, when it will make your life better? Put your pride to one side, do not swallow it, it will only make you constipated with the sickly sweet aroma of narrow-mindedness.

88. A non-record:
We have been unconscious of what as been going on, or we have been out of it, for over 200 years, with wholesale efforts made by the upper echelon of society to mask their own folly, and point the finger at the poor unfortunates who seek nothing but immediate gratification.

These upper echelon people have even falsified historical records to hide their own shameful involvement in the risks to the planet. They have watered down their own culpability and passed on the blame to their descendants, leaving them to deal with the ever-growing problem of global warming.

89. The Co-op shop in the desert:
The biggest 'shop' in the world, the Australian Co-operative in the desert has never been recognised, but it is there.
There are no signs, in fact there are no buildings, nevertheless, you can 'shop' in any 'department' and obtain whatever it is you need; shelter, food, water, or just a sense of wellbeing, none of which are available in tins, packets, or freezer bags. Furthermore, unbelievably, these goods and services are free.

Author's Extra Note: Sounds too good to be true doesn't it, but it is true, all can be had for nowt, and that 'price' has not changed in over 80,000 years.

90. Mechanical mayhem:
As in many other countries, the private motorcar in Australia has been in the forefront of social advancement, making distance less of an obstruction, and keeping communities in touch.

In all fairness, taking distance and space into account, the harm done by car exhausts on the Australian continent must rate as one of the lowest in the world per capita.

However, allowing this ad hoc method of calculating, the effect on global warming, is not just weak and foolish, it is downright obnoxious, in more ways than one.
You may have a gleaming comfortable car, but it will probably be one that is helping to kill the planet. Will you get rid of your shiny new car? Not bloody likely, the future is not your problem, leave it to your children's children and their children sort it out; and why should Australia do any different to other countries? Let the mayhem continue.

91. Sweeping a road in Sydney:
It might be a trivial matter to concern ourselves with, (if you think so, go read a comic book) but when a mechanical road sweeper moves along, it is more than likely creating more invisible damage filth that it is cleaning up.
The simple commonsense approach would be to employ a person with a wooden broom. It will take longer, but it will be global friendly, and help point a way forward for many municipal scientific engineering academics, and the cost would be a lot less in financial terms as well, what you might call 'a no brainer'. We have risked killing the planet with a 'box of

matches', now we can help save it with a 'witches broom', fantastic. Come on give me a break.

92. Windmills - stone towers with sails:
All is no lost, there are remedies to hand that will not cost anyone a smidgen of pride or much money for that matter. All around Australia there are stones of all sizes, and people to lift and carry them, with respect.
Constructing a Dutch type windmill will not take much effort, pile a few stones, cement them together and you have the base of a windmill.
A few hundred years ago, the Dutch had an excellent way to deal with encroaching seawater- they built hundreds of windmills and lifted huge quantities of the water, to make it work for them, and not against them. The same opportunity is now presenting itself to Australia, not to move seawater, but to replenish the dry hinterland with fresh river water.

93. The Dutch Polder and its windmills:
Polder, a word for reclaimed land, but whereas the Dutch word was for land reclaimed from water, with the use of water pumps powered by windmills, the Australian windmill system is for the opposite, to reclaim lakes of freshwater out in the bush and in the outback.

Freshwater lakes surrounded by a green and pleasant land, a true panorama of the future, and one that can be had by the introduction of the Dutch windmill.

94. A vast sea of sheep:

In the dark of night, the 'sea' is calm with just a slight breeze disturbing the setting. The sheep are still and quiet, with no predator to bother them in the dark. However, in the day before, the sheep's predation on the grass beneath their feet had been forthright and relentless.

At first light, the sea of sheep begins to ripple, and sounds of bleating swim through the already warming environment.

The ripple changes so quickly, as the sheep move in a steady moving tide, and the day's grazing begins in earnest, and will continue for most of the daylight hours.

Behind the huge flock, the land is has changed to a degraded barren waste, while in front of the moving tide of sheep, the land is so huge and verdant, that the threat posed by the sheep seems to be negligent, with seemingly no indication of the devastation to come, as come it will.

95. Topography:

Much of Southern Australia consists of a series of gargantuan basins created by the continent's slumping and faulting of millions of years ago.

By the way:

96. Western Australia:

Covering roughly one third of Australia, with less than one tenth of the population, Western Australia is an expansive region of natural beauty.

Great Deserts dominate the interior, tall spring flowers flourish in the southwest; white sandy beaches and rugged cliffs rim the coast and extraordinary rock formations rise from the untamed wilderness of the Kimberley in the far north, and it can be very hot.

Noteworthy:

97. Truth:

Real honest indisputable truth of a good thing or event is most wonderful, and you can depend on it. However, the truth of a bad thing or event can only be unpleasant, or as some pundits say 'hard to swallow', and neither good news or bad news can be denied.

98. Walking on wet sand:

To walk on wet sand is not something you might worry about, until you begin to wonder why it is wet.

99. Right, she'll be right:

Being right is a cheerful condition, and being in the right place in the world at the same time, makes being right absolutely bloody marvellous. That's what being an Australian feels like.

By the way:

100. Highways and Byways:
Almost all Australian highways are still just two lanes wide, and this brings the countryside, or 'bush', that much closer. All the details of the landscape are right there with you, up close, and while you are there, your arse is part of it to.

Chapter 7. Desert:

101. Sand above water:

The desert is bountiful and teaming with life in the good seasons. It is like a vast untended communal garden, the closest thing to earthly paradise I can imagine, but a wee bit lacking in garden sheds.

03 October 2018.

102. Desert 1:

The frost can cling like brittle cobwebs to the bushes around, while the sky will turn thick with glitter, and then the sun is switched on in the morning, and all is aglow again.

103. Desert 2:

Vast plains covered in yellow grasses, like wheat fields swept up to the foot of chocolate-brown rocky mountains and ranges.

104. Desert 3:

In every direction, for as far as the eye could see the earth was covered with spinifex, a brittle grass, which grew clumps so closely packed as to give an appearance of greenness.
'It looked like the land could support a thousand head of cattle an acre. In fact, spinifex is useless with needle sharp points tipped in silica.

If the land grows plants, why not they be plants that can be sown and harvested? Maybe that is too complicated, or too much of a politically and cultural 'hot potato'.

105. The Great Victorian Desert:
Most of The Great Victorian Desert, which is on the border of South Australia and Western Australia, lies within the Officer Basin.

23 October 2018

106. The Nullarbor Plain:
Along the sweep of the south coast of Australia, there is the Nullarbor Plain, a crescent of limestone almost undeviating in elevation and covering some 250,000 square kilometres.
The Nullarbor was once underwater, today it is treeless and over water. To the north are dune fields with dunes that are the characteristics of the plain.

107. The Simpson Desert:
You can take an aircraft and fly over the endless wastes of the Simpson Desert, and not think of walking across it. However, some have done just that, with the first difficulty being to begin the walk. On the other hand, the beginning might have been the beginning of something good.

I hope that modern techniques will overcome the original explorers' lack of knowledge and technical assistance, and the beginning will become less daunting.

22 December 2018

108. Desert 4:

The land around so far as could be seen, was mainly flat , with the tufts of spinifex, low bushes, some rocks that have been there for thousands of years, and above it all a clear blue sky. If you are Aborigine what's not to like.

109. Sandhills:

Hills made up of loose granular particles smaller than gravel, and coarser than silt. The structure can be of any size and height, and is always moving.

110. Making money in the Desert:

Why? Where are you going to spend it, and if you save it who benefits?

You take no more from the desert than you need, this way you will leave enough for those who will follow you, like those who went in front of you left for you. No profit or loss, but loads of benefit.

111. The Desert for Families:

The desert areas, are where many explorers/settlers sometimes died, was, and some folk still do in some cases, but at the same time it is where large families of Aborigines lived, or live, in adequate and simple comfort. Managing the land by water control in times of flood, setting fire to dry growth at the right time, and planting crops and harvesting, and then storing the produce in more agreeable periods. The fact that those families survived and still do is testament to their grasp on how we should live.

Author's Extra Note: It is somewhat absurd to think that a box of matches could save the planet.

112. Actions and effect:

Whether we take action or not, we all have an effect on our environment, and each and every one us are involved. Maybe we can learn from all that is around us and beneath our feet, no matter where we are, maybe.

113. Water in the desert:

An old man with young ideas, and a deeper understanding, carried a sealed bag of freshwater into the desert, and when he arrived at what he considered the middle, he let the ground near his feet feel the water. He then sat down and waited to die.

Ten million other old men did the same, and the old man, the desert, and the planet, were saved. Could it be that bloody easy? Of course, if you are an intellect or scientifically knowledgeable, the previously mentioned will be a load of twaddle, but, remember, so called 'twaddle' has saved folk before, and it can do again, and in a very big way

114. Flying:

In the early days of flight, a journey by air from Britain to Australia would have taken up to twelve days, and a lot of courage and luck, and possibly more than one aircraft. Is this where we would use the phrase 'a wing and a prayer'?

Chapter 8. Survival:

115. Survival in the Outback:

If you were to be missing in the outback, you would be as lost as a white rabbit in the wilderness of the local supermarket, and your tenure in the world would be counted in days rather than months, or maybe only hours.

116. Survival 1:

An escapee survived a week on roots and wild nettles until the Aborigines caught him and gave him to an armed search party of constables.

117. Sustainability:

'The land was not arcadia; the bush could flare up and incinerate ten years of a forty-acre-man's work in a day, even in good times it took three acres to sustain one sheep.'

118. Generosity and openness:

Generosity and openness of Australians in a tight spot does exist, just don't be a 'grindle' of a berk and suck on it - if you are o.k. say so and 'buggle' off.

119. Survival 2:

Turning desert sand into water and food - a seemingly mad definition, a non-starter you might say, but for over 60,000 years the aborigines have done just that, and on a large landmass-scale

Author's Extra Note: Being alive but only just, is a phenomenon we have to accept. We do not think about it as we make our way through each day, whereas in Samuel's case, his reason for 'just existing' is all around him.

120. Distance:

Imagine a world with no precise measurements, or a need for them, no precise measurements of distance required. A world where distances are calculated in an ad hoc fashion; 'close by', 'near', 'far away', and 'don't bother'. How calming is that? It is Aborigine-calming, that's what it is.

20 January 2019

121. My Kimberley Watering Hole:

The Leslie H. Harvey of Kimberley Award.

    The Leslie H. Harvey of Kimberley Service Award

        to

      Greggs at Kimberley

'A Takeaway with Seating and Civility'
*****

Frequented by Leslie H. Harvey of Kimberley
To be awarded on the 23rd of May 2025

122. Survival 3:

The capacity for survival is not an unusual trait for humans or animals, or plants for that matter. But whereas animals and plants change for the better in order to survive, the human section of this planet's occupants seems to ignore such a mundane longwinded apparition, and will go for the quick remedial solution, regardless of cost to others or the environment. Like it or not, or agree or not, we are on a slow decline along with the planet, one we are ignoring at our peril, and Mother Nature will react to this in the only way she knows how, and we will not like it, but we may well have to lump it if we leave it much longer. So, get of your hacienda and do something about it, make a difference now, and you will not need a massive bank balance to do so.

Author's Extra Note: To be in a state of isolating poverty, is to be in a privileged position, but only if isolation is considered to be a higher state of being that is.

123. Truth getting in the way:

Some may fantasise of what the real Australian is, but truth-gates are in the way and as mere mortals we become becalmed, and 'thwaddled'* in media gobbledegook. *A Les H Harvey/Aborigine word of many meanings.

124. Enemies:

To radiate discomfort and even hatred, you need look no further than your enemies. In order to eradicate this threat you can make a friend of your enemy, but you must first make sure you can stop being an enemy to them.

125. Racial superiority my hacienda:

Once upon a time:

The Incas thought they were superior.

The Greeks thought they were invincible.

The Egyptians could not find fault in how they lived.

The Romans conquered all and expanded their empire.

How or where are they now?

The civilisations of today, including the one you are part of, will they last?

What is more important, will the planet last?

Author's Extra Note: While long ago civilisations were coming and going, the Aborigines of Australia stood their ground magnificently, and they are still there.

126. Expectations of Travel:

To travel in the hope of arriving is one of humankind's main characteristics, and the Aborigine people were no different. However, maybe we should think of how they did it, and why. In the very beginning of the populating of Australia, the land was pristine to the touch of a human hand, and to the human eye, there were no predation of the animals, not as we know that to be today.

The whole country was at peace with itself, everything in balance. The land prospered, and what humans and animals were there at that time, belonged there, and had learned to survive, and not destroy.

It is not too late to get the collective brain in gear and learn again, and very soon please.

Chapter 9. Global Warming Part 1:

127. Wet and Dry:

In Perth, the dry shift has wrought major impacts on water systems. Between 1997 and 2005, area stream-flow dropped 30% below those observed in the previous 23 years, which were themselves on the dry side. In 2002-06 Perth launched Australia's first major desalination plant in the industrial town of Kwinana, 25 km (16 mile) south of Perth., the largest such plant in the Southern Hemisphere. It uses reverse osmosis to render water from the Indian Ocean drinkable.

The plant will supply an estimated 17% of Perth's water needs, although that might solve the shortage it is powered by a climate friendly wind farm.

127. (a)

The ozone hole is above Southern Australia, and the rainfall across the southwest has declined over the last half century with a sharp decline in the mid 1970s. Since then, rainfall in Perth has been between 10-15% lower than before.

127. (b)

As of 2005, Perth has gone 38 years without reaching 39" of rain (100 centimetres) per annum for 1.5 million people in the Perth district. In simplistic terms, this equates to 1 mm of rain to share between 150 people per year. With trillions of 'freshwater' escaping into the Indian Ocean every day down the Swan River, were it then becomes saline and then desalinated at Kwinana.

14 November 2018.

128. The Drying of Southwest Australia:

Water- and the lack of it - are key elements in the psyche of Australia. The unofficial national poem, Dorothea MacKellar's *My Country*, extols a land "of droughts, and flooding rains." As a severe dry spell raged in 1888, Australia's Henry Lawson bemoaned his fate in verse: "Beaten back in sad dejection, after years of weary toil/ On that burning hot selection where the drought has gouged his spoil."

128 (a)

With much of the continent at the mercy of wild swings in precipitation, many Australians have viewed the southwestern coastal belt, roughly from Perth to Leeuwin, as a corner of relative climate sanity. One of the world's bio diverse areas, it features Mediterranean climate with hot, dry summers and mild dependably damp winters. A 1920 book called *Australian*

Meteorology noted the region's reliable moisture: "Here the rains rarely vary 10% from their average amount, and the lot of the farmer should be a happy one."

128 (b)

The farmers-and the urban water managers-haven't been so happy lately. Rainfall across southwest Australia has declined notably over the last half-century, with an especially sharp shift downwards in the mid 1970s. Since that point, wet season rainfall has constantly run some 10-15% lower than before, most noticeably in the late autumn and early winter. These days a wet year in Perth is one that merely reaches the long-term average of 869mm (34.2"). As of 2005, the city had gone 38 years without mustering a yearly total of 1000mm (39"), a mark once reached regularly. For a booming city with 1.5 million thirsty people, those are very scary statistics.

128. (c)

How much of this is related top greenhouse gases? Moreover, will it continue? We must remember that much of the countryside was deforested after European settlement, and computer models show that the resulting landscape evaporates less moisture into the air, perhaps exacerbating drought.

128. (d)

The last few decades of drying do bear some of the fingerprints of global warming. Scientists have verified a pole ward shift in the storm track that girdles the Southern Ocean, encircling Antarctica. The low-pressure centres that race eastward along this track are the source of nearly all of Perth's rainfall, and they also provide important cool-season rains in Melbourne and Sydney.

128. (e)

Computer models agree that the storms tracks winter position is likely to shift even further south as the century unfolds. This could mean real trouble for Perth and vicinity, with a 2 degree centigrade warming, which could produce a 20% reduction in rainfall across much of Southern Australia.

We could expect degradation of ecosystems, considerable loss of agricultural capacity, and severe water restrictions on southern cities. The cost could run into billions of Australian dollars, and the action would be like using a plastic frying pan.

128. (f)

Wintertime drying has also been noticed in other more eastern places such as Sydney, where substantial amounts of summer rain in the form of showers and storms make it more likely to survive the global warming effects. But even these rains have scaled off in the last few decades and the future could be a lot dryer than is comfortable.

128. (g)

In Perth, the dry shift has already wrought major impacts on water systems. Between 1997 and 2005, area stream-flows dropped by 30% below those of the previous 25 years, which were themselves on the dry side.

21 November 2018.

129. National Climate.

Australia is a long way off, on the other side of the world from most other places.

It may be a misconception of mine, but I when I think of Australia, my thoughts are of a hot dry outback, red scorching sand, and flies.

However, as I write this, the temperature is Sydney in the day is 27 and 14 at night. In Alice Springs, it is 35 by day and 24 at night.

130. Australian Climate Council and Global Warming:

The Council called for critical action as records tumble, but was anyone listening, and if so would they be doing anything about it?

Autumn brought no relief to the situation following a record-breaking summer, a summer driven to high and dry weather conditions by rapid global warming.

Author's Extra Note: If you are content to ignore the existence of global warming, then might I politely suggest; if we do nothing about it, that you arrange to come back one hundred years from now, and then see if you can still ignore it.

131. Deaf ears:
Deaf ears 'listen' to warnings that Australia is not on track to meet its agreed climate target, does this mean that the political lobby are fully aware, and are taking on board the real danger posed by global warming, or are we being fobbed off by political jargon yet again, while the planet dies? Now let's be clear about this!

132. Why worry:
We should be aware; or we should be *made* aware, that we could and should change our way of living now, and care more for each other, and the planet that is our home, before it is too late. Quoting reasons why we should turn a blind eye and a deaf ear are of no real significance when the sea levels rise by metres and not millimetres, by then it will be too late. However, it would seem that the powers-that- be don't give a shittle anyway. As long as there is money in the bank, why worry.

133. Saving the Planet by saving Australia first:

While the world at large is populated by folk who seem to be incredibly oblivious to the demise they are perpetuating on Planet Earth, and don't seem to have a blood clue as to what to do, the folk of Australia have the key to a much better and securer future.

Author's Extra Note: Claiming that international political agreements are needed to make the 'saving' possible is just turning the blowtorch, and avoiding difficult but very necessary actions.

International politics and politicians are of no use, it and they only serve one master or mistress, single-minded commercialism and its profit motive. However, If Australia leads by a different and more nature friendly example, the rest will follow.

134. Delusion and Diffusion:

Across the huge land mass of Australia, there are folk living out their lives in somewhat difficult, if not appalling circumstances, but doing so successfully, and not all are Aborigine.

It is these folk who can lead the world away from the potential destruction that; in the guise of modern living, the majority of us seem to be striving for.

To defuse the highly regarded anti 'green' situation, and cure the delusion the human populace of the so-called developed world seem to admire, is far from an easy option, but it is of a bloody important one nonetheless.

Author's Extra Note: I know it is stating the obvious, but this Planet is the only chance we have, lose it and we lose everything, including the future. We are all individuals, but collectively we are one human race, 'creatures' with the ability to make things safe.

135. The Oxford Book of Sydney. 1788-1997:

The enigma of 'listening' to dead people talking, and maybe even learning something important, it's there for all to see, in a book, or on a computer monitor, or just by looking and listening - it ain't rocket science, and it could save the planet.

136. After 2018:

Just as the Aborigines did before 1788, the modern world should manipulate and manage the new changing environment of our planet before the worst happens, and not wait until it is too late. It is not out of the question that in order to produce food aplenty, we can live where we grow the crops.
This is a tall and somewhat seemingly stupid hypothesis, but we have to do something - and quick in terms of a new and natural evolution. If we continue to misuse the planet it will fight back,

and we won't like it. Tough shuttle you might think, but that is how Mother Nature works, tooth-and-claw and all that.

137. White Supremacy: (Even thinking that leaves a horrible taste in my mouth)
Supremacy of race, a fallacy some modern folk consider undisputable, yet it is 'we', the supreme modern human community of whatever colour that is destroying the planet. The phrase 'white supremacy' was dominant in reports of the early explorers and settlers of Australia in the 18th and 19th centuries, their words would not give any literal evidence to the 'native' abilities, and this engineered a disbelief and ignorance in the aboriginal doctrine. However, they completely ignored the lifestyle, the way of living in harmony with nature, and consequently opened up the pathway that is now taking us towards destruction. If you are on the bandwagon, my advice is to get off, or continue to do what the three monkeys do.

By the way:

138. Darwin on a hot day.
You do not need to be in the desert to feel the heat, but there are the cushions of white buildings with verandas, louvered windows, potted palms, lazy ceiling fans, cool drinks in tall glasses, men in white suits and Panama hats, ladies in floral-print cotton dresses and floppy bonnets. c1920

Chapter 10. Disabled:

2 September 2018

139. Travel Finders:

When on a plane take a seat at the back. You get a longer ride, and you can usually be close to the 'lift' for you and your scooter.

14 October 2018.

140. Getting to Australia:

Threadbare thoughts on how to get to Australia: It is cheap and easy to imagine, in fact, it is free, but will only achieve an imaginary journey made in the mind and on paper. However, the reality is more expensive, even with a new Mobility Scooter. Maybe things will change and the imagining will become a reality for many.

141. Disabled Travellers: NICAN

(National Information Communication Awareness Network) This is a good place to begin, although Sydney caters reasonable well for people with disabilities, it is wise to start making enquiries before leaving home - steps from one level to another can be a problem.

A good place to begin is the (NICAN), a national organisation that keeps a database of facilities and services with disabled access, including accommodation and tourist sights. It also keeps track of the range of publications on the subject. See NICAN. tel: 1800 806 769; www.nican.com.au

142. Disabled Travellers: IDEAS

IDEAS (Information on Disability Awareness and Education Services) also offer online databases on disability services, equipment suppliers and accessible travel, plus other speciality information for other agencies. IDEAS, tel: 1800 029 904; www.ideas.org.au.

Extra note: We who are disabled are in a minority, and, like it or not, we are treated as such. Ain't life a peach when it is first come first served.

143. Metro Monorail and LightRail

These services are wheelchair-accessible, as are some buses (indicated on timetables available at bustops or online at www.sydneybuses.info). From my personal experience, the wheelchair- accessibility bit does not include mobility scooters, unless its one your can carry in your back pocket or purse.

By the way:

144. Manly Ferry:

(Sydney) Catch a ferry from Circular Quay to Manly (30 minutes). In Manly amble along the Corso, look in the shops and then enjoy a Sydney experience with fish and chips at Manly Fish Market, while sitting on a bench near the Manly Beach. https::// www.myfastferry.com.au

Chapter 11. Home Australia:

145. Close by:
If Australia is your home, you are extremely fortunate. You may only be a mini speck in the social mix, but all around you are folk who know what freedom means; in a country that stands out far above the other more 'plebish' populations of this world.

146. The ground beneath your feet:
Take a look at the ground you are standing on, it's been there for thousands if not millions of years, and you are exercising the privilege to stand on it.
You can build on it, make fires on it, flood it with water, grow food on it, and it will remain in situ for you to repeat ad infinitum, and this as been going on for a length of time we cannot imagine.
Now think of it in a different way, in a way that gives some credence to the fact that you understand the privilege you are exercising, a privilege the Aborigine women have always understood. You are in partnership with that land beneath your feet, don't be a pillock, instead embrace the fact.

147. The Australian Way:

A blasé attitude of 'I don't give a shuttle just won't work anymore, the continent of Australia and the world are in too much danger, It would probably better to instigate the Australian attitude of 'I don't give a stuff for hypocrisy or bigotry.' But we need to be aware of the political fraternity, they are about as much good as a fart in an ocean of flood water. Better that we adopt the Aborigine society's calm and easy approach, that way we will be on the right track. They have had 80,000 years of experience, when are we going to take heed? But then, we can consider ourselves superior to such easy going folk, can't we, and make even more mess, it would seem that we are really good at it.

148. Singing Australia:

'Australia Fair' is a statement of reality, and it is best we listen, not to listen would be discourteous.

Take away the trappings of nationalism and we see whom we really are, Earth folk with every one of us in a partnership of survival, not just of the fittest, but of the need to ensure we and our descendants of the future are and will be safe.

21st of January 2018

149. The right way and the future:
Who is to say which is the right way for the future? Not yours truly that's for sure. However, let us assume that we do eventually do the right thing, what then?

First of all, we get rid of the niggling doubt that no one seems to want to qualify, or in some cases not even signify, and that is the undeniable fact that our planet is getting warmer, or is suffering 'global warming' if you would dare indicate that such a phenomenon dose exist.

Jump forward by two hundred years or so, and imagine Melbourne consisting of sand dunes and an inner city wild bush status, with dry waterholes and scrubby tree growth where there are now green parks and gardens. Not a pretty thought is it, yet it could come about eventually. However, I have faith in the human psyche, that logic will rule by honour and justice, and the land will become safely 'Aboriginalised', and guarded by a new non-political indispensable commonsense fraternity.

150. Pillocks: (Hindsight is a wonderful thing)
I think it safe to say that some of the original European settlers in Australia where pillocks of the first order. They did not or could not see the sense in adapting the Aboriginal way of treating Australia with a respect based on commonsense, and, as a result, there as been two hundred or so years of full-scale neglect of a beautiful country. Now is the time to sit up - oh

bowgger that - now is the time to *stand up* for Australia, or allow the *new* pillocks to make another mess of things.

151. Anonymity:
The fragrance of a free thought can drift through your mind either in little wafts, or with a full-blown overpowering scent of realisation, and the rumour of a good and easygoing honest lifestyle we could have free is a good thought to ponder on.

But anonymity means the chance of a new good lifestyle is buried under a huge pile of bureaucratic gainsay of ignorance, and the chance of that good life slips way as it is made to fit political boxes, to remain lost forever in obscure legislation. .

152. An interfering hacienda of a Pom:
To Mr and Mrs Australia: Yes, okay, 'an interfering hacienda of a Pom' is a good description of me, and if things weren't as they are, I would accept the title. However, the prevailing circumstances mean I will have to forego the pleasure of being known as such, and robustly point out that my feelings about it are of no importance, and any thoughts you might have of applying the title to me are of no importance either, not when you seriously consider my reason for being 'an interfering Pom'.
All around the world people are thinking and saying that global warming is bad for the planet, and others are saying it's a load

of bullshuttle, including one Australian politician who shall remain nameless, but they are doing sweet buggle all about it. However, there is a chance for us. The Australian people are in a fantastic, if precarious position to take things more seriously and more positively.

What is happening in Australia now is the foretaste of what will happen to the rest of us, but Australians are in an unenviable lead, and whether they accept the challenge or not might hinge on whether the planet survives or not. God bless all Australians.

153. The Great Australian Peace:

Stupid wars have taken their toll, and like other countries, Australia has needs to remember.

However, the continent of Australia is one at peace, a land that all can consider as a contented, although somewhat racially aggravated homeland.

I will try to be delicate, but, my dear reader, thousands of men women and children have lost their lives in what the politicians call a sacrifice for peace. Then, as soon as the dust settles, they're at it again, pointing fingers like uppity school kids, showing evidence of the injustice done by one group on another, and instead of sitting down and sorting it out they proclaim and show a sign of strength; my bomb's bigger than yours, etc. This is just loads of bullshuttle, and we all know it is, but no one seems to give a gnat's whatsit for peaceful settlement. It is as if non-aggression is boring.

Author's Extra Note: For the bad to succeed it only takes the good to do nothing.

Come on Australia, you've got a head start on the rest of us, show us the way, or just sit back on your hacienda and resolve to be like the pillocks the rest of us have been, and in some cases still are.

154. Non-Australian Civility. As apposed to Australian commonsense:

It is probably not wise to assume that because of higher education, the western society toffs are better able to solve Australia's problem of Aborigines versus t'others, not when the Native Aborigines have proved beyond doubt their ability to care for Australia successfully. It is good to consider that they organised and cared for the whole country before 26th of January 1788, and really well it would seem, and for over 80,000 years.

Western or 'white' minds are not superior; they just contain a lot of non-essential garbage, which doesn't leave much room for rational thought patterns. Is being pedantic a western habit, or is it an excepted trait of any of us? Even so, I can't believe that an Aborigine would be so ped-ant-ic, not even if he or she knew what that word meant.

155. Rivalry:
What's all the guff about Melbourne versus Sydney? Get bloody real, make it Melbourne *and* Sydney, and while you're at it, why not make it Brisbane *and* Cairns? Ah, but then, we have to consider the politics of it all don't we.

156. Ferry:
Watching the ferry to Manly leaving in an unhurried but seriously busy fashion, on a journey that will be repeated many times over, and one that had been done many times before. It is like the ticking of a clock on water, and one that folks never grow weary of looking at.

157. Smells:
The smell of scrubland and desert in the dry is different from that of in the wet, but it is still scrub and sand, dry or wet, and it is still Australia.

158. Reaction:
This is not a purely Australian problem, but it is my belief that like it or not, Australia is on the front line, or, if you like, first in the queue for disasters.
In our present mode of couldn't give shit, we are not equipped to put right the damage we have done and are continuing to do, to Mother Earth, and in time, what I believe to be a short time,

the planet will react in a very fierce way, and Australia is on the back foot.

By the way:

159. Nobody likes a smartarse:

Try as I might, I cannot think of myself as a smartarse, I would much rather regard my thoughts and actions as fair, and distinguished by my efforts to make life better for all of us, and not just the 'smart set'.

If you do think of me as a smartarse, you are at least thinking of me. Just take it a step further and think of what I am trying to achieve, and don't be a smartarse yourself.

Chapter 12. Politics:

Author's Extra Note: It might be said that I have an axe to grind here; not true , I just don't like the two faced lying 'doggoes' who call themselves politicians-for-the-people, and act like the two faced money grabbing self-ingratiating-toadying-thieves they are.

160. Bureaucrats know and do nothing about the Aboriginal way of life.

They are either too bliidy scared or they are too bliidy thick, an acumination usually defined by the single word 'political'. If it's a lie, it's more than likely political.

6 January 2019

161. Power and Opportunity:

The imbalance of power and opportunity, and most importantly, freedom: we need the power and freedom, and the opportunity, to make up our own minds outside of any political or material influences, and if there is an imbalance its bloody wrong.

Every breath we take is a promise of the future. We can make the opportunity, and we can manage the freedom, and, what do you know, we can have influence, every single one of us.

162. Intelligentsia and the Planet:

Are the political intelligentsia intelligent? If so, why is the planet in such a blood mess?

Of course they are intelligent, but only in a political or academic self-preservation way.

They cannot stand up and declare their frailties, their inadequacies, or their unwillingness to actually do anything about planet warming. They bundle about with facts and figures, and do nowt (nothing) that means anything to the rest of us, nor for the people who will inhabit this planet in the future, what might turn out to be the 'dead future'.

163. Divisions:

There are no political divisions that cannot be made good by formulating a political agenda, but when this is done the politician seeks fame and fortune, and not the idyllic result of political manoeuvring for the good of all.

However, and please take this into account, where there are politicians there is a plentiful supply of bull fertiliser, fertiliser that grows nothing but inadequacy and downright hypocritical conduct such as 'Let's be clear about this'. When I hear a politician use that phrase I inwardly cringe. Being clear means, or should mean, being honest, and I ain't known of an honest politician since Abraham Lincoln, and even that's based on historical hand-me-down hearsay.

164. Indifferent People:
Living with an absence of interest or sense of importance in the environment, and ensuring that non- important matters are dealt with succinctly, by keeping them away in order to belittle their importance, is maybe not good practice. If that's gobbledegook, then so is the modern outlook on lifestyle-versus-commonsense, and survival of the planet.

165. Buggle and Boob:
Sorry to be so ped-ant-ic, but, in my book, a politician is about as much good as a plastic fire shovel, and they're not even a necessary evil, they are just a waste of space. Not politically correct? Oh buggle it, I've boobed again.

166. Make Australia politically free:
Make Australia safer, get rid of politicians and turn desert into lush green pasture, and then we are on the right track to saving the world, but involve the politicians and literary perfectionists and you might as well sit on your thumbs. They will arrange meetings and discussions on how to disagree, and then look around for someone to blame when the shuttle hits the fan. Ignore the 'polly tissons' and make good their past mistakes, lift the ban on straight honest thinking and actions, and save sweet Australia and the wonderful planet we call home. Perhaps

take a look at Bingara and Alice Springs and start the recovery there.

167. Administrations and the Politician:
Serving the community by either shuffling unnecessary paperwork or sleeping on a bench/seat in a parliamentary debating session. These qualifying misnomers are part of the political life of any individual who purports to be a member of parliament, or some other governing body, a servant of

the public my hacienda. Ten out of every ten of 'em are in it for what they can get out of it for themselves. Sacrifice the many to serve, sorry, *save* the few, eh?
The paper-shuffler will leave in the evening, and come back in the morning to shuffle the same bloody paperwork, with a job satisfaction that goes as far as I can throw pig trough full of swill.
On awakening, the slumbering politician will shout approval or otherwise of a speech of which he or she has not heard one word. In this way, the political 'ornaments' will fulfil their obligation to vote for or against a motion,* the bearing of his or her vote having been decided before the speaker of the speech uttered a word.
* Normally just a load of obnoxious or worthless waste. (How's that for diplomacy?)

168. Those in favour:

Months are taken to compile a list, but the list cannot be completed for political reasons.

169. Those against:

Months are taken to compile a list, at great cost, but the list cannot be completed for political reasons.

170. Those undecided:

Months are taken to compile a list, at great cost, but the list cannot be completed for political reasons.

171. Supreme back scratching:

There is no such thing as racial supremacy, it is just that one race of people are different from another in back scratching. Within one race of people, there will be those who really matter, and those who find it judicious or enriching to scratch a back or two, the consequence of this being a sense of wellbeing by living cleanly and shrewdly.

Those folk that matter will get by in harmony, good fortune and good recollections in old age, attributes that can be worth more than any amount of back scratching will provide.

172. The top hat and brolly brigade:
Rules and regulations have to be made to protect those who are vulnerable, and it is the Civil Service 'wallers' who make ready, to notify and implement compulsive legislation, and see that the poor unfortunate folk are herded into bracketed files. This top hat and brolly regime is managed by civil servants who are eager to be in vogue, always ready to obey the politicians, to see that laws are implemented and acted on with outright unflinching accuracy, no matter who gets hurt in the process.

Author's Extra Note: Please do not confuse 'twaddle' with political waffle; waffle is speaking in a vague and wordy manner, sometimes over a prolonged period, whereas twaddle is definite, although maybe pretentious talk.

173. Political Global Warming: Or to put it another way, 'the idiot's guide to destroying a planet.'
It is my considered opinion, for what it's bliidy worth, that Central Australia is where the first hammer blow of the real global warming syndrome on a dying planet will fall, if it hasn't already. When it does, the erstwhile sceptics will jump forward, and proclaim that 'we should have done something about it earlier, before things got as bad as they are now.' My dear sceptics, things are as bad as they can be already; get set to make that jump forward now. What are you waiting for, a political proclamation from the planet Mars?

By the way:

174. Aussie slang: (Put your tongue in your cheek as you read, but don't try and speak).
'Mate'. Pronounced 'mite'.
'Pom'. British, possibly English.
I did not include any others, as offence might have been looked on as offensive, then I thought; 'is a Western Oriental Gentlemen better at catching rabbits than a Pom'?

175. Key resources:
In an idyllic world all that is required for good honest living are good air quality, fertile soil, clean water, and good decent folk to share them with.

Communicate:

176 The Royal Society of South Australia:
Can I communicate with such dignified folk and make a difference? Can they communicate with each other and make a difference? If they communicated with you would you listen and take heed?

Author's Extra Note: We of the so-called modern and intelligent societies should be in awe of the Aborigine, but racial bigotry gets in the way, and we move along like a snowball heading for hell.

Chapter 13. Global Warming Part 2.

01 January 2019:

177. Long-sighted:

A new year as begun, and I am in hope that a new Australia is rising above the scholastic indifference that has got us to where we are, up the old creek without a paddle, with the stench of political engineered manure all around us. You can be sure that politicians will have answers for that, along with the phrase, 'let me be clear about this'. All politicians are steadfast in assuring their own future, no matter what the cost to others: let *us* be clear about that!

Author's Extra Note: It would seem that the answer to the warming of our global 'village' problem, could be the introduction of a totally integrated system, in which all constituent parts are interdependent and linked. In a nutshell, we need to get our collective act together in a global way.

178. Cooling the Planet:

Plant and grow trees over the vastness of the Australian hinterland, and these will draw done enough carbon dioxide (CO) from the atmosphere to eventually help chill the planet. A childlike answer to a global problem maybe, but it does have a modicum of good sense.

As irritating as it might be to accept the Aborigines had it right for over 80,000 years, all we need do is catch up. Get off the ivory tower of smugness, get real, and help save the planet, even if it is in a childlike way.

Alternatively, we could continue to think and say we of a worthier intellect know best, as the planet dies under our arses. What an overwhelming concept:

>Your back leg is on fire but you knew best.
>Because you were more intelligent than the rest.
>The planet is failing under your feet.
>Your hypothesis is incomplete.

179. Pussyfooting:
Treading or moving wearily, or stealthily, to avoid committing oneself to a course of action. Are you a 'pussyfooter', or are you eager and willing to do your bit to save the planet?

With Sydney Harbour extended inland by two hundred kilometres, many Australians thought how fortunate they were to have a vast desert world to set their sights on. My eyes are on the distant prospect of having surfing lessons on the beach at Bingara. Every silver lining has a cloud, or has it been wrongly suggested by those of little understanding.

180. The Butler's day awf:

Without doubt, we all lack commonsense at times, and then regret it. Now the human race is in a bliidy tricky position, and only commonsense will get us out of it. At the present, I don't give a monkey's left ear about protocol, or political correctness, and I most certainly do not countenance paying heed to so called experts: it is such as they who got us here in the first place.

Rid yourself of any thoughts of political programmes or initiated cooperation between that bunch of hypocrites, those pretending to be possessing of virtues, and that bunch of hapless experts. They only give thought to what's in it for them, even while they are proclaiming agreement on global warming. It's like an Aborigine blowing a didgeridoo, breathing in through the nose while blowing out with the mouth, or to put it another way, political blowing of one's trumpet when it must be the butler's day awf.

181. The end is nigh: (Where have I heard that before)
'Let's cut it short, don't hang about; go to the end of it all, and we may well be doing ourselves a favour'. The words of a 21st century sceptic, words our descendants will learn off by heart, muttering them as the planet begins to disintegrate beneath their feet and all around them, with the seawater rippling along the inland valleys that were once prime areas for development of easy living communities.

The hand held computer gadget in their hand will be of no use, and their eyes will fill with tears of woeful disappointment.

There will be millions, if not billions, knowing at last that we got it wrong, even though the Aborigines had a way out for us, we chose to stay our pride.

You can waffle on about bullshit and other like attributes to a warning that will hit home, but one that you will undoubtedly bloody ignore; it is more socially comfortable to do so.
Your waffling is indicative of a race of beings bereft of commonsense, what the Romans, Greeks, and Incas did, we are about to repeat, but on a massive scale. If there is any comfort to be had, it is the thought that though we are not the first to balls things up, we may be the last, but, and here is a mind blowing bloody idiotic consideration if you like, we of the 21st century may be the last to cock things up big time.

Authors Extra Note: Please allow me to point out that there are some animal species that no longer exist, we humans have seen to that, and we continue to take comfort from dangerous unhealthy living while we make more animals extinct.

182. Single-minded degradation:

'Untold damage', a two-word phrase often used to circumspect a situation, brevity in the concept of a world in the process of being killed by its inhabitants. To use this kind of phraseology to highlight a rate of degradation is about as ludicrous as we could get.

This planet, our home, is a robust spatial object of a massive size, and we are, and have been fortunate to have such a home. Humanity has progressed materially and physiologically over the centuries, and, up to now, the planet has surprised us only on the odd occasion.

However, sadly and seriously, the old home planet of ours is beginning to suffer greatly, and like the living thing that it is, it will react fiercely in order to survive, and much more often, don't say you have not been warned.

183. An epitaph to the ignorant:

Who will be there to write an epitaph to human frailty and stupidity? I who wrote the question will be long gone, and some of the people who have read the words will have gone the same way.

Whomever it will be at the end to write of our stupidity will have the problem of being fair to us folk of the 20th and 21st centuries, we who uncaringly got the unstoppable planet-destroying ball rolling.

184. A cry in the night:

A cry in the night of an animal, a kangaroo maybe, trapped and unable to free itself. What do we do? Ninety-nine times out of the hundred, we do absolutely bloody nothing.

Like that kangaroo, our dear home-planet is trapped and unable to break free. What do we do? We play political lip service to our collective non-caring nature, and do next to nowt.

However, unlike the poor trapped animal that unmercifully died a slow agonising death, our planet will react and turn the tables on us, and it is we who will die the slow agonising death. What a 'lovely' smutty thought.

185. The rule of Mother Nature:

Mother Nature lives in Australia like she lives all over the planet, and she has her own rules, and no matter what we do to break those rules to get somewhere we would like to be, we will find ourselves struggling.

We are now in the process of breaking the ultimate rule of Mother Nature by ignoring the climate change we are perpetrating, and one that is taking place right in front of us every bloody day and night.

Slowly, and most certainly, we are beginning to face the wrath of Mother Nature, and we will lose.

186. A new future for our world:
Grasp the nettle and ignore the stings, make no tentative moves, be sure of what you are about to do, so sure that the consideration of adverse results makes you even more determined.
The way back is going to be a long life-changing route to take, and there is no room for doubt. We need to be sure, not in apolitical sense 'to do what is best', we know that doesn't work, but in a social interaction that brings about a natural response that includes Mother Nature in a full partnership.

187. The new road ahead:
Any half measures will not do, neither will smart-alecky notions be worthy of consideration. If we are all together no matter what race, religion, or social standing we are, then the new road will be more open and free, and the world will be a happy and safe place.

188. Getting it wrong again:
When the European settlers got it wrong in the 18th and 19th centuries, the Aborigines paid a very harsh price. If we get it wrong now, it will be all of us who will take the consequences of a massive planet-ending price, if 'price' could ever be the correct word to use that is.
There will be no following populations to hide their shame, and say sorry.

189. Final reckoning:
190, 191 and **192** and the planet:
All dead?

193. The last address:
In all probability, it will be the Aborigines who will be the last to go; they are more resilient and know Mother Nature more intimately than the rest of us. They will not get it wrong, but, through no fault of their own, the Planet Earth will be their last address.

194. Greed versus Pride:
We are all familiar with the phrase 'pride comes before a fall', but when greed precedes pride, or it seems to do in odd places in Australia, the fall may be a long time coming, although, and think about this, it will come eventually, and it will be a huge global fall, with cataclysmic results hard to imagine.
If you are to be greedy, keep it within sensible parameters. Be proud, but not at the cost of the lives of millions around you. Let commonsense overrule greed and pride at all times, and the future might be at less risk.

195. Battle against the tide:
As I write, in the year 2019, the sea is many distances away from me, and I have no worries about the tide overwhelming my position.
Of course, I can concern myself with what I envisage to be a definite actuality, but I think my words will be superfluous, of no real value.

However, being the old stubborn old fart I can be, I am looking forward across the many distances of desert and stony ground that is inland Australia. Nevertheless, my mind is ultra clear on the fact that global warming *will* 'lift' the level of the seas, enabling the morning tide to move ever closer to places hitherto hundreds if not thousands of distances from the shoreline of today. Bingara-by-the-sea will look great on a picture-email.

196. A wasted greeting:
A new dawn of a new day, and the rising sun is there to greet all who care to pay heed. However, in the busy business of the morning, the greeting goes unnoticed and is wasted; allowed to go unobserved.
How many more greetings will there be to go unobserved?

By the way:

197. Blame:

To apportion blame is to judge, and I ain't no judge. The whys and wherefores of the Aborigine people's national decampments, and demise in some instances, practically by overnight enforcement, are not for me to judge in a factual way. I can only be aware that injustices have been enacted and I have no right to make any suggestions. It is up to the people on the scene as it were, and the instigators are long gone, and unanswerable to any questions others or I may feel the need to ask.

Author's Extra Note: I spent some long time on this section, taking part in open discussion and examining my innermost thoughts, with the result I felt it important enough to deal with it on its own, in a more investigative way. I sincerely hope you can understand my dilemma, and my heartfelt concerns. In all honesty, I am not fully satisfied with my effort on this, but hopefully you can grasp what I am about.

Chapter 14. Aborigine 1:

198. Understanding the Aborigine sticky situation:

Author's Extra Note: If you can't be bothered, it's alright, we'll manage without you.

Aboriginal:

1. Spiritual communication with the land.

2. Withdrawal of Aboriginal cultivation systems may have had a fierce detrimental outcome on the Australian outback or bush land, early settlers, new Australians, hadn't got a clue. What there is there now is 'artificial', and nothing like what it should and could be, and once was.

3. All behaviour is an expression of a well-developed sense of moral conduct (which provides the basis for all human imperatives). See 'Dark Elm' by Bruce Pascoe.

4. We are responsible for our planet's health, and not the other way round, muck about with Mother Nature and she will slap your wrist bloody good and hard.

5. The spiritual and social dimensions of technology are of greater importance than the actual practical application.

6. Tools derived from wood or stone are loaded with ethical and spiritual obligations and significances, you don't get that from a spanner bought from Joe Bloggs' market stall.

7. Early reports of Aboriginal history of dramatic and noteworthy development are culturally and socially suspect. The early 'journalist' didn't know his head from his hacienda when it came to Aborigines.

8. Mother Nature and Mother Earth can give an excellent supply of good food if treated with respect.

9. Environmental destruction in NSW - when the primary degradation of land and extinction of mammals occurred - was due to the introduction of sheep and British farming methods. It wasn't a case of poor soil, it was a case of unfortunate farming methods.

10. Australia is different, and now is the time to celebrate and explore that difference by showing the rest of the world how to look sharp, and make this planet a good and safe place to live, the original Australian way.

11. The greatest difference between the culture of Aboriginal Australia and that of mainstream Australia is the concept of land, with the New Australians operating on the basis of what's yours is mine and what's mine is my own.
You can have a mocking whinge about that, but if the cap fits, be good enough to wear it.

12. Nature, 'Mother Nature', encourages health and balance, and the spirit of beauty within each of us, and all the food we need, to radiate with a healthy glow like the morning light or the sky at dusk. In other words, by getting of our collective beam end, and knowing bigotry for what it is, we can make our world safe, and more.

13. 'Humbled by the most ancient, bony, awesome landscape on the face of the earth'.
Australia is massive, its 'bush lands' are empty to an idle observer, but full of potential to any one with an inkling of what the Aborigines are about.

14. The red dust was choking me, the blasted heat was roasting me, and the Australian fly population seemed to be interested in me alone, but I wheeled on in my 'wheelabout'.

15. 'The endless wastes of the Simpson Desert ... the lush green sogginess of coastal country'. Combine the two and make Australia green and great again.

16. 'Desert; purity, fire, air, hot wind, space, sun, desert, desert, desert,' and more desert.
Then along comes the Future Freda's and Fred's with 'don't care what you think'.
They put water where there is no water, and plants where there are no plants, and what do you know, a green Australia and the chance of a better and safer world.

17. Stop here for a moment, stop and think. How do we make the right moves for a safe future? Maybe *how* that is important, it is more that we think of the people of the future. It is my belief that without a political delegation or political divergence, we can and will make the right moves, we just have to, there is no excuse.

18. The modern Aboriginal people now have to contend with the complex physical, political and emotional problems in their everyday life, syndromes that were brought on by non-caring early settlers, who, like it or not, looked upon them as didgeridoo playing natives of no consequence, and that is putting it mildly.

19. Alice Springs: Aboriginal Legal Aid
Aboriginal Health Service
Aboriginal Arts and Crafts Centre

Department of Aboriginal Affairs 'DAA'.

27 September 2018
199. Utopia:
Utopia is a 170-square-mile cattle property, which had been 'given' over to the Aboriginal people.
The country was flat, grassy, covered in tall scrub in places, doted with lakes, and through it ran the Sandover River, an enormous white sandy bed that swelled to a red raging torrent *when* the rains came.

28 September 2018
200. Diet in the Desert:
Even in the good season, I could prefer my diet to be supplemented by the occasional tin of sardines, and a frequent cup of sweet tea, and an ice cream would be fantastic.

201. A rock is part of a net (place).
To you it's just a rock, or just a stone, but if you could allow yourself to be less intolerant, you could take in the real

significance of what it is, and get a whiff of real Aboriginal utopian lifestyle, and it won't cost you a cent or a penny.

202. Gordon Wayne:

An artist of aborigine artwork. Panels consisting entirely of romantic and idealised landscapes, showing herds of kangaroos sipping at a billabong, or swagmen gathered around a lonely coolibah tree. My own version is what I call a Story Stick or Story Wood.

20 October 2018

203. The Mabo Case:

The Mabo Case was a form of legal representation and acceptance of Aborigine Rights that even today are not implemented completely, just give it time: said the constipated politician.

204. The Merion People:

The rights of the Merion people make them the traditional owners of the Murray Islands (including the islands of Mer, Dauer and Waier)

205. Aborigine Association:

AIATSIS

GPO BOX 553

Canberra

ACT 2601

Australia

This is an aborigine association, and I would like to be a member if possible, but that will probably be in the future.
https:// www.mycommunitydirectory.com.au

'It is a great delight to dwell in a picture.'
D H Lawrence (1885-1930)

206. Aborigine reading:

Scholar and Sceptic Australian Aboriginal Studies in Honour of LR Hiatt: January 1997
by Francesca Merlan, John Morton and Alun Rumsey ISBN 855752955

Australia: by William Blandowskis, Harry Allen and others.

Country of the Heart: Deborah Rose and others.

The Power of Knowledge, The Resonance of Tradition: by Luke Taylor and others.

The Little Yellow Red Book: Aborigine Indigenous Australia.

207. Resettlement:

When the Aborigine people were dismissed from Ooldea, and the desert nearby, (the 'Whites' call it Nullarbor) they were forced to settle permanently on the unfamiliar, un-storied limestone plain to the south. The Aborigines then found that they became ill on what they called *Pana tjilpi* or 'grey earth'. They had no resources to combat this devastation, and thousands did die, and the poor native people gradually gave in to the inevitable, and they are still doing so today. What a bloody legacy.

Then came even more disastrous manoeuvres by the white population. Once the atomic tests in the 1950s began, the people were not allowed to go back, not until the 1980s, when the Oak Valley aboriginal settlement was established Oak Valley is 100 kilometres north of Ooldea, along a dirt road over the sandhills. A sign reads 'Warning no alcohol/no drugs on the Maralinga Tjarutja Lands.' Political Masonic manoeuvring at its bloody useless best; if your face fits it's your shout.

'At Biak, the ground glittered like a lake that had just begun to freeze, and it crunched underfoot as I walked. The stuff encrusting the surface was trinitite, named after the place of its first earthly creation, the Trinity nuclear test site in New Mexico. The grassy green substance occurred when the sand was fused by the heat if the bombs'. Oh, at least the future will look green.

208. The Arangu Tribe:

The Arangu, like all aborigines, are a people who can preserve a narrative that has been unchanged for tens of thousands of years. The English language as it is today could have been four thousand years in the making, whoopee bloody do. To put into some kind of perspective, if we use the human maturity rate as a guide, the English language is at the 'mammy-gurgle daddy-gurgle' stage when compared to the old Aborigine languages.

209. Oak Valley:

Four hundred kilometres to the northwest of Ceduna there is the Aboriginal community of Oak Valley.

There are plenty of Aboriginal places where it is forbidden for non-aboriginal to go, a kind of crude and cruel equality me thinks.

210. Undertone of Aborigine:

'Ten hours and 903 kilometres after leaving Daly Waters, we arrived in Alice Springs, a grid of streets set out on a plain beside the golden slopes of the MacDonnell Ranges.'

Sometimes golden with a touch of blood hot, and cold, with an undertone of Aborigine.

24 October 2018

211. The 400-year-old Aborigines:

At the beginning of the 19th century, it was thought that Aborigines had been on the continent for more than 400 years.

212. The 20,000-year-old Aborigines:

In the early 20th century, it was estimated their time in Australia to be around 20,000 years. then...

213. The 23,000 year old Aborigines:

The skeleton of an aborigine woman was discovered in the nineteen sixties, and the carbon dating
indicated a time span of 23,000 years. then...

03 November 2018.

214. Aborigine 'Life-Stick.':

What I call my Aborigine 'Life-Stick' is just a piece of old wood, but by my associating it with what I have learned about the Aborigines, it has become my connection to those who where in Australia prior to the 18th century.

The 'stick' is a piece of bark from an old tree that must have started growing approximately at the end of the 18th century, around the time of the first settlers arrival in Australia. It sits on my desk at all times, and is in front of me now as I write, a symbol of my new life at ease with Mother Nature, and I am much better for it. Life is so uncomplicated; all you need is a stick

06 November 2018.

215. Pencil and Paper:

Have you got a pencil and paper? If I asked that question of Aborigine Samuel in the old days, his answer would probably have been a smile and a shake of his head. A simple illustration of life in an Aborigine settlement, where things we take for granted are not even thought about.

Even if he did have a pencil and paper he would probably have had no use for them, life was simpler that way, with aural traditions that committed huge amounts to memory, sometimes in song and dance.

08 November 2018.

216. Fire in the Outback:

Fire, grass, kangaroos, and human inhabitants, they all seem to depend on each other for coexistence in Australia.

For any one of these to be wanting, then others could no longer continue... The Aborigine applied fire to the grass at certain seasons in order that a young green crop may subsequently spring up, and so attract animals, and enable him or her to kill or take animals.

In summer, the burning of the long grass also disclosed vermin, birds' nests etc., on which the women and children, who are involved in burning the grass, feed.

217. Aborigine: (In the old days maybe)

On rising each morning an Aborigine must find and catch their own food, make or repair their own tools and shelter, and defend and educate their families. They are their own provider, manufacturer and protector, and there is no supermarket needed.

218. Pitjatjarna People:

Many of the old folk may not speak or understand English, while the people on the whole have managed to keep their cultural integrity intact, in any language.

Pitjanjara: '*Nyuntunpalya nyinanyi Uaw, palyaran, palu nyuntu*'

By the way:

219. Food.Vitamin 'C':

Solancceae is a huge family, including potatoes, tomatoes, capsicums, datura and nightshade. Many of these form a staple diet for Aboriginal people, a source of vitamin 'C' when their enforced 'western' diet is almost devoid of vitamin 'C', giving rise to crippling health problems.

30 September 2018.

220. Plants:

'A new plant would appear, and I would recognise it immediately because I could perceive its association with other plants and animals in the overall pattern, its place in an Aborigine Society you might say.'

  Anon. c1890

221. Staple Diets:

Brown rice, lentils, garlic, spices, oil, pancakes made with all manner of cereals, and coconut and dried eggs. Various root vegetables cooked on the coals, cocoa, tea, sugar, honey, powdered milk, and every now and then, a can of sardines, some pepperoni and Kraft cheese, a tin of fruit, and an orange or lemon.

Supplement with vitamin pills, wild oats and some meat. It just depends on which Supermarket you go to, and/or where you are.

10 December 2018.

222. The Store:

'The store was typical, a small galvanised shed, selling the basics - tea, sugar, flour, the occasional fruits and vegetables, soft drinks, clothes, and billycans. It was refurbished with goods once every couple of weeks by road-train or light aircraft

from Alice Springs, or maybe even by a bowlegged camel with a gleam in his or her eye'.
Anon. c1952

Chapter 15. Railways:

223. The Indian Pacific Railway:
From Sydney to Perth. Runs for 2,270 miles, through New South Wales, South Australia and Western Australia.
From Sydney it climbs gently through the Blue Mountains, then across miles of big-sky sheep country. It then traces the Darling River to the Murray River and along with the Murray, and finally across the Nullabor Plain to Kalgoorlie and Perth. It's along way to go for a night out.

15 October 2018
224. On the Indian Pacific: (At 100 kilometres per hour - my Mobi' Scooter as never gone so fast)
On the Indian Pacific: 'The first class section consisted of five sleeping-carriages, a dining carriage and a plush velvety style lounge bar, with soft chairs, and a small promising looking bar with low but relentless piped music: it was time for dinner and the tannoy announced the first sitting.

225. The Nullarbor (Latin 'No trees'):
For hundreds of miles the landscape is as flat as a calm sea, and barren, just glowing red soil, clumpy tussocks of bluebush and spinifex, and scattered rocks in and off a sort of brown colour,

with no shade. It is one of the most forbidding expanses on earth, but there is water if you know where to look.

226. In the Driving Cab of the Pacific:
Snug and comfy, with a fancy console complete with lots of switches and toggles, three shortwave radios and two computer screens, and a number of domestic comforts: electric hotplate, a kettle, and a small refrigerator.

227 Reaching the Sea:
Indian Pacific Length of route: 2,704 miles (4,352 kms) or 432,640* Double Decker London buses, that would be one super traffic jam. *If you want to arrive at a more exact figure, go for it.
Indian Pacific Running Time: about *four days! (In England that would be in hours, in any direction before reaching the sea)*
Stops at: Perth. Kalgoorlie. Cook. Adelaide. broken Hill. Sydney.

7 January 2019
228. The Ghan:
Length of track: 1,851 miles (2,979 klms)
Running Time: 84 hours (Not in one go)
Stops at: Darwin. Katherine. Alice Springs. Adelaide.
On wheels that normally go round on a steel track way.

The train was called the Afghan Express in its early days, named from the Afghan camel traders who made the first commercial 'white' routes through Australia, south/north.

8 January 2019

229. Puffin Billy:

An old railway, and engine of the late 18th century, and making a vintage run through the Dandenong Ranges.

It takes 90 minutes over a distance of 15 miles (24 klms), between Belgrade and Glenbrook in Victoria. The downhill sections are possibly a wee bit faster.

230. Spirit of Queensland:

A railway of North Queensland, a line on the east coast. Brisbane to Cairns, with contact to the Great Barrier Reef.

Built in: 1924

Length: 1,044 Miles (1,680 klms)

Times: Five trains per day.

Author's Extra Note: To travel young is to educate, when to travel old is an experience.

By the way:

231. Shangri-La Hotel:
Shangri-La Hotel.
176 Cumberland Street
The Rocks Sydney
NSW
https://www.etrip.net/shangrilahotel/sydney
A$280 per night per room for two. (Prior to 2018) Cumberland Street runs beneath the Cahill Expressway, and the hotel is across the Sydney cove from the Sydney Opera House, and is close to if not beneath the Sydney Harbour Bridge. The Shangri-La Hotel is comfortable in amenities and is ideally situated. Most large hotels- apart from the more moderately priced ones – have lifts. This hotel has rooms for the disabled.

232 Park Hyatt Hotel:
Park Hyatt Hotel.
7 Hickson Road
The Rocks
Sydney
NSW
Australia
www.sydney.parkhyatt.com

Chapter 16. Just a matter of years:

233. The shadow of memory:
You can see yourself in your memory, see and understand the wrong moves you made, and when and what the effect was on your life.
However, if you make a bad choice, or make a wrong decision, you are obliged to both forget it and move on, or you can remind yourself constantly, and live under its shadow. It is your choice.

234. Doing right:
The obligation to do right by others is not to be taken lightly; even animals must feel obliged to be right with others in an instinctive move.
Extra Note: I have witnessed this myself with domestic cats, one cat helping another cat to give birth to her kitten.

235. Australian by Proxy:
I could do the literary equivalent of running of at the mouth, but I will try to desist, and keep to the matter at hand.
When my pen strokes the paper, Australia and Australians are paramount on my mind, and I feel a compelling need to write

about such a good country and its good people. The need to write about both is probably because, right or wrong, I consider myself Australian by proxy, programmed to think Australian and try to enjoy life in the Australian way.

There you are, I said I wouldn't 'run of at the mouth'. God bless Australia and all Australians.

236. Have a go:
There are plenty of blank spaces in this work for you to make your own comment... feel free.

237. 80,000 years of being right:
A whole race of people had been right for over 80,000 years, but in the last two hundred or so years however, the fact of them being right was nullified by a newer race of people who were more sure of being right in a miss-aligned devout way.

Using their religious devotion, and their conviction that they were superior, these new folk made life near on impossible for the original more down to earth natives, by using the excuse of converting those natives to a better way of living. The fact that the natives were right, and had been for thousands of years was swept away, and the devout ones declared themselves to be right by invariably doing wrong.

238. Innocence versus Ignorance:

Whereas 'Innocence' can be an excuse for lack of knowledge, or judgment, or even just lack of commonsense, 'Ignorance' is maybe just a 'get out' clause.

In any event, 'Innocence' or Ignorance' are of no use in any attempts at bringing a true nature- friendly future to this planet of ours. There is no room for second-guessing or smudgy thinking; we need to be sure of what we are about, what we are aiming for.

If either Innocence or Ignorance are allowed to rule, the result will be of no use to the planet, and, in a natural consequence, the demise of all flora and fauna will lead the way. Is there any sign of this natural demise? Yes, there is, and no one is taking a blind bit of notice, or if they are, they are pleading innocence by ignorance, in a naive political gesture of doing what is right by doing wrong, or nothing at all. What a bloody stinking mess.

239. Enough stone to build a city:

Assuming that all has been done to turn back the stupid tide of ignorance, and what was once an undeniably seemingly human weird desire to look after number one, and nollocks to the planet, has been halted.

On the vast continent of Australia, there are stones of all sizes, and rock faces that will easily provide the same, stones that can be used for the construction of all kinds of buildings.

With the previously mentioned in mind, if Bingara is a small town, then New Bingara can be a city; there are sufficient raw materials to make it so, all that is required is the power to achieve it. The same kind of power that built Sydney and Melbourne for instance can rise again.

Author's Extra Note: When using rocks and stones check with the Aborigines as to their ethical availability.

By the way:

240. Odd Places:
'Dog Swamp'.
'Inner Loo'. (Inna Loo?)
'Sunbury'.
c1836: In Sunbury, Victoria, in 1836, settlers, including Isaac Batey and Edward Page, observed that the Aborigine 'people' had worked their gardens so well and for so long that large earthen mounds had been created - in other words there was good soil management.

Author's Extra Note: Proof that cultivation was a feature of Aboriginal land use.

Chapter17. Sydney:

22 October 2018

241. Sydney 1. Harbour Bridge 1:

It is difficult to keep focussed when dealing with day-to-day matters that include expenses, but I have not lost sight of the Harbour Bridge in Sydney – there is a photograph by my desk, along with Uluru, Melbourne Railway Station on Flinders Street, Koala Bears, Kangaroo, and a map of Australia. What is not by my desk is the names of the good men who lost their lives while at work on building Sydney Harbour Bridge, and that is something not often quoted, until now.

29 October 2018

242. Sydney 2:

Sydney continues to grow has a city, but it remains close to nature. Near the city centre, on the fringes of the harbour foreshore, near Rose Bay and Manly, you can find yourself ensconced in a bush landscape that the First Fleet folk would have recognised.

27 November 2018.

243. Sydney 3: Australia's largest, oldest and best-known city is located on the shores of a great natural harbour, and basks under a mostly sunny sky, in a mild climate, a symmetry that beckons to residents and visitors alike.

30 November 2018

244. Sydney 4. Harbour: The Sydney harbour on a sunny day, with its blue waters, is usually complete with ferries, yachts, cruisers and ocean liners.

Author's Extra Note: It may be just an area of water, but it is as played an immense part in many lives, with a bridge that spans it in a kind dignity, with the harbour matching the bridge with its own kindness.

245. Sydney Harbour Bridge 2:
Opened in 1932.
Length: 503 metres (1,650 ft)
Crown of the Arch: 135 metres (443 ft)
Width of Deck: 49 metres (160 ft)

8 Motor Traffic Lanes.
2 Railway Tracks.
1 Cycleway.
1 Footpath.

3 December 2018

246. Sydney 5:

Even the tourists have a yen to be part of Sydney's daily business; they will even walk over the top of a bridge to be so.

7 December 2018

247. Sydney 6. Harbour:

'Sydney Harbour is 9 fathoms deep, it is 7 miles from open sea, with very little tide, and with the sweetest weather found of earth'. c1803

Noteworthy:

3 January 2019

248. Bourke 1:
Hot! Even on a 'scoot'.

15 February 2019

249. Bourke 2:
Still hot! Scoot or not.

Global Warming Part 3.

250. Fight Back:

Our biggest friend is the planet we live on, yet it seems we treat it with contempt. It is a stupendously wonderful creation that consists mainly of water. If we keep on abusing it, make no mistake, it will fight back, and in a bloody big way.

Dreaming:

21 February 2019
251. The Dreaming:
The Dreaming implies a code of conduct, a form of behaviour, and a pattern of life. It implies active custodianship (of land, of scared sites, of relationship and of people) as well as acceptance of doctrinal rules, the 'Dreamtime' is of tremendous importance, and it is where earthly reality meets the truth of spiritual significance. Look in a mirror, turn three times and look again. If you feel dizzy, screw yourself the other way.

252. Keeping the dream alive:
Dreams can be cherished, if not submitted to the degradation of the historic 'couldn't give a hoot' syndrome.
The non-Aborigine folk who are not party to the 'dreaming', should probably consider themselves unworthy, devoid of any sense of a natural kindred, and fully at risk to be known as the killers of the planet, or at the very least those who began to engineer its death throws of fire and flood.
Keep the dream alive or else.

Chapter 18. Aborigine Part 2:

253. Aborigine temperaments in nature:
There is a peculiar power and strength in the country, which in many ways expresses itself in Aboriginal people and which I feel can belong to me too. It keeps unfolding and unfolding and is inexhaustible. What you make of it depends on you, no matter where you are, or who you are.

27 February 2019.

254. Aborigine Salvation:
Many were slaughtered, with the remainder forced to live on settlements that were more like concentration camps. Then they were poked, prodded, measured and taped, and had photos of their sacred business printed in colour. They had their secret objects stolen and taken to museums, had their potency and integrity drained from them at every opportunity, and have been reviled and misunderstood by almost every white person in Australia. Then finally, they were left to rot with their cheap booze and the diseases given them by white folk, and then their subsequent early death from those diseases. It is time to return things to the better, to the salvation for everyone.

3 March 2019.

255. The Aborigines Caretakers:

The Aborigines must be one of, if not the oldest settlers on this planet, and it must be beyond doubt that they are the original settlers and caretakers of Australia.

256. Aborigine Huts for Homes:

Huts made with arches of strong tree branches and covered with a good coating of clay. These buildings were a good projection against wind and heat, they were of a good size to accommodate a family or more than one family, and they cost nowt (nothing'), they were mortgage free. All this was, and can again be covered by a balanced good for all way of life.

257. Aborigine Home Village:

Even in the harshest parts of Australia, the Aborigines made their homes, and managed the land by digging wells, growing crops and/or harvesting the wild plants for food and shelter, and using controlled fire.

258. Self Destruct Button.

It is there for all of us to recognise, the historic evidence, and the proof that what the Aboriginal managed to do was right. Be derisory by all means, but remember when it comes to shovelling with a shovel, the best with the shovel will survive, especially when pressing a button won't mean a thing other than destruction by neglect.

259. Sturt and 'Encampment':
'The whole encampment (village), with the long line of fires, looked exceedingly pretty, and by eleven in the evening all was still.'

260. Aborigines and Endurance:
Aborigines did not cling on as it were; they thrived in an environment the early white explorers and settlers thought of as harsh, but what the native Aborigine considered verdant enough to provide for a joyful life, the 'white folk' just could not see it.

261. Sacred Ancestors:
Dreamtime stories of Aboriginal ancestors who roamed Australia thousands of years ago are sacred. They indicate and personalise the creation of the lakes, hills, caves and rocks, with Uluru being a sacrosanct example revered by the Anangu folk. Between these features, there are paths that are 'songlines' that still exist today.

(Author's Note: I considered it a real privilege for me to write those words).

262. Pitjantjara 2:

This is a genuine Australian native family, or tribe, with a fifty to sixty thousand year old dynasty behind it. Nowhere else on earth is there such a living dynasty, a dynasty that is part of a greater native expanse of folk, people truly of the land that is their home, and they have looked after it for all those thousand's of years, and looked after it well. Give 'em bloody credit will yer!

263. Partnership with the land:

The way the Aborigines live may be distasteful to most non-Aborigines, but if we think on, think of the thousands of years they have been successful in raising families, cultivated the desert land by being in partnership with it, and not needed any false wealth or materials. If we can be blood honest, they must have something we have not, something we now desperately need.

264. Desert Foods:

There is food in the desert if you know where and how to look. It might look barren but it is not. An Aborigine will know where and how to look for food, and water, and there will be no sense of panic
in the looking.

265. Time 1:

'The Aborigines do not have much time. They are dying'. A quote (Not her own) from 'Tracks' (Page 119) by Robyn Davidson c1980. Is this true today? I sincerely hope not.

266. Time 2:

The original concept, of around two hundred years ago, and through to modern times, was that the Aborigines would gradually die off, and leave Australia to the new settlers, who were mainly European with no perception of any dynasty of a native empire. But it would seem only the Europeans knew what was best for Australia. It is only the dogged lifestyle of the Australians that has heightened that shortsighted attitude, to a point. Sadly, there are bits of that kind of old-fashioned bull fertiliser still around.

267. Time 3:

The Aborigines have been in Australia for over a 1,000 of our lifetime span, and, like it or not, they are senior to today's population by around 80,000 years.

They are a population that has lived; survived, for all that time, without the help of modern thinking or adapting to modern ways. It is a European enforced shift life protocols that has nearly killed off a dynasty, and there is no pride to be had from that, none whatsoever.

12 March 2019

268. Time 4:
In the time the Aborigines have been successful living a good life in Australia, the Roman Empire has gone, the Greeks have been of no consequence for a very long time, and the British Empire is no

269. Nankari:
An Aboriginal Doctor of great wisdom, with an aptitude that far exceeds the modern concepts of medicine, will have no need for Latin words or terminology, which serves no one but the medical profession, keeping its quasi methods adrift from commonsense common knowledge, and at much risk to the sufferers. The Aborigine Doctor will cure with knowledge of healing from thousands of years of keeping the Aborigine population safe and well.

270. A lack of water:
Samuel wakes up to another dry hot day, with only water and food on his mind; there is no room in his thinking for prosaic matters. The promise of death, no matter how disguised, must be a soul destroying flagrant misuse of the gift of optimism.

271. The Aborigine Women:
Over thousands of years, the Aborigine women have had the power to bring new life into the world; and they have the

knowledge of how to live off the plants and creatures of the bush and desert.

The women know the meaning of the Aborigine ceremonies, and how to perform them. It is these ceremonies that are important to the continuance of the force of nature in balance with their lives, and it is the women who allow the men to glean some knowledge from such ceremonies, and then use the information for the betterment of all.

272. The Aborigine Creed:
If there is one Aborigine Creed, it is that of the true concern and partnership for and with the land.

I do not profess this to be academic in anyway, and my approach to matters is not what might be termed as scholarly. However, on my own way through life, I have had times of association with the land beneath my feet, and it is these associations, simple and not erudite, that have brought me to within a gnat's whatsit of understanding the Aborigine way of life. For me to take a glimpse at their outlook, and their involvement with our planet, is to see with fresh eyes. If there is a way to save this planet, the Aborigines have it, in abundance; after all, they have had 80,000 years of experience.

15 March 2019

273. Indigenous People:
Living naturally in a particular region or environment, and taking care to make all that is simple important. The opposite lifestyle, the 'modern' one, is too complicated and imperfect to the extent of being blood dangerous, but it is acceptable, and it is the norm because it is profitable, and sociably 'nice'.

By the way:

274. The Middle of Australia:
On the first of April, 1860, John McDonnell Stuart reached the dry centre of Australia - halfway between Daly Waters and Alive Springs, 'the most desolate place to be'.

Author's Extra Note: To stand a chance of knowing the real Australia, watch the sunrise in the morning, and watch it go down in the evening, and realise that you have been privileged, to witness such an eventful day.

Chapter 19. Alice Spring:

275. Alice Springs 1: (46 degrees of heat and plus).

Alice Springs was constructed on parched red desert dust, and is the original town of the Australian pioneering spirit.

Founded as an Overland Telegraph Station in the 1870s, the rough and ready town has evolved into a popular destination for tourists.

276. Todd River:

A 'scattering' of aborigines were strolling along the dried bed of the Todd River in Alice Springs. They had no place to go and no need to go there.

277. Alice Springs 2:

There are over ten thousand people living in Alice Springs, of whom 1,000 are Aboriginal.

There are three pubs, a few motels, a couple of medium grade restaurants, and shops that sell T-shirts and boomerangs made in Taiwan, and various books on Australia.

It is a frontier town, with mannish ethics and severe racial tension, but you can say 'good morning' and mean it.

.

278. Alice Springs 3:
There is an old building in Alice Springs that looks as if it belongs, what is a dilapidated old stone house nestled in the hills. Some time ago, someone had a dream.

By the way:

279. The Way of the Gumtrees:
'We were in a creek-bed where tall gum trees and delicate acacias cast deep shadows, cool deep shadows.
Everywhere there are birds: Black Cockatoos, Swallows, Major Mitchells, Willy-wagtails, Quarrian, Kestrels, Budgerigar flocks, Bronze wings, and Finches.

There are also various Solanums and Mulga apples, with eucalyptus 'fruit' to eat as we walked along'.

Noteworthy:

280. Sydney to Perth:
The train journey from Sydney to Perth – takes nearly three days. Start at Sydney's Great Central Station on Elizabeth Street, then Flemington, auburn, Parramatta, Doonside, Rootty Hill, and the Blue Mountains, and head out west.

Chapter 20. Expectations Part 1:

21 March 2019

281. The Future:

We must regard with great importance the future of the world, and our judgment should not be weakened by socially directed economic requirements.

282. A new unity:

'I believe that humankind can find a new unity through a love for humanity and nature, independent of the need for monetary gain or of puffed-up prestige.

283. The Modern Predicament:

It is all well and good basing our predicament on a kind of profit and loss account, but if it is profit or loss, without a correct natural involvement it will matter not one spittle. We should remember that life is important to all creatures and plants, without it the bacterial fart becomes king or queen, until the bacteria spread their wings and claws or whatever, and wave goodbye to competition. To put it into context, whoopee-do for the bacterial lifestyle, by whatever name, Covid or not.

284. 'Foot Shooting':
If we continue to be modern and 'shoot ourselves in the foot', then the future looks very bleak. The so-called intelligent folk of today will experience, and probably deny to the very last, the same fate of the early explorers and settlers of Australia, and learn the blood hard way, and in a retrospective affiliation, experience the same downfall the Aborigines have had to contend with for more than two hundred years.

285. Good intentions:
The road behind us was paved with good intentions, and brought us here to a new year of a mixture of hope and despair. The hope is that Australia can lead the world, has it did before, some thousands of years ago, while the despair is that we failed to listen two hundred years ago, and are maybe doing so again.

I have good intentions and the words here before you are my pavement. If the authorities, the politicians with fingers in the wrong place, can see sense in my words, and in the performances evident in Aboriginal history, then maybe, just maybe mind, the intended road ahead will be a good, safe and solid one, and paved with more than good intentions.

286. Intelligentsia:
'The intellectuals form an artistic, social, or political class or vanguard.' A dictionary definition of these folk is profound and

succinct, but is it viable in these extraordinary times? Is it possible to be intelligent and foolish at the same time? Ask the folk who venture out into the vast arid areas of Australia and expect their intellectual knowledge to save them.

If the cap fits blooming wear it, but don't expect the wrong cap to fit every circumstance. You can't dig for water with a computer, you might find where it is but you will be dead from thirst before you get to the life preserving liquid, and your computer battery will have run out anyway.

4 April 2019

287. Australia can and will show the way:

Australia is in a unique position in the evaluation of human designation as planet dwellers.

The country is beset with problems no other nation has to contend with and it does so adequately, with no bother about the rest of the worlds clam footed living regimes that make those nations fully blameworthy* of killing the planet. *I don't like the word, but why should I pussyfoot around?

There is room in Australia to make changes that will affect the planet in a good way, making rivers become resistant to drought, and making cities more relevant in the true sense of being part of an already warm country.

To show the way forward is not a degradation of human rights nor is it likely to denigrate the Australian life, in fact the exact opposite is very likely. Making Australia great in terms of saving the planet can only be the best Australia has ever achieved, and the way forward could in fact be the Aborigine way, by taking pride in Australian history retrospectively beyond the 18th century in retrospect.

Author's Extra Note: It is 5.35 in the morning, and I am taking my first coffee of this day April 4th 2019. My medicine I am sure is keeping me alive as I sit at my desk writing. I wonder if my Aborigine friend Samuel is out of bed in his village, and is he taking his medicine?
That question made me reflect on my good fortune, thinking of how I make it day to day on medication. My good wishes go to Samuel and all those in his village, and I hope he and they are okay.

288. Taking a step back to potential:
More than 250 or so years ago, Australia was free of any cumbersome modernity, life was free for all inhabitants and life was good for those who just lived without the barrier of commercialism and 'modern' lifestyles centred in technological advances.
If taking a step back is grasping a nettle, then grasp the bloody nettle and get on with it, stop reminiscing on how things were. Move forward with a backward step into a real conclusive life, life that we can be proud of, a community we can be proud to

be a part of, a world wide web of decency, of rightful togetherness, to live in harmony with the planet and each other. We can of course put it in the political jargon: let it be stated here and now that Australia as a nation can be proud of its achievements. On the other hand, to be 'politically' correct, make that 'some of its achievements'.

289. Winners and losers:
For the 'don't give a shuttle' folk to win, the good clean folk just need to do nothing. The folk in the first category will happily allow this planet to die, as long as they can get there way, while the folk in the second category are just too apathetic to care, apparently.

290. Fresh Australia:
Blood in ell, can I justify what I am going to write here. I am not qualified to be politically correct, neither am I qualified to be socially correct, for that matter there are not many that are, but I will give it a good go.

This planet we are living on is dying, expiring at an ever increasing rate, and the hierarchy, whether they be scientifically academic or political, appear to be doing absolutely nothing with any real quality that will give this planet a chance.

Okay, the expiration may take hundreds of years, but it is beginning now, and it is up to us of today to change direction before it is too blood late.

I could tiptoe around, and accept the blithering idiotic gainsay service these people are offering, and be part of their band of plonkers. However, that's not my way. If I think it needs to be said, I will bloody well say it, and as loud as I possibly can, and with the tread of a bloody elephant if it I think it will work.

A New Australia maybe this planet's only hope. If we can make it in Australia by changing the systematic bull-it of the hierarchy, and mature with the land around, then we will be on the right track, and without any clever echoing gainsay advice from learned people who think that only their valuable contribution is worthy of recognition. Recognition my arsenic! If you are Australian, thank you for being Australian. You have an inbuilt gift you may not recognise, but I can assure you do have it. Just stop listening to such as me and be Australian, please, the world needs you. The planet of ours, our home, needs you.

291. Small World:

As time passes by, this large planet of ours, if left alone, will survive by changing as per requirements in a completely easy and calm fashion. It is only the less significant human world that will cause it to make massive and destructive alterations in a reactive unsteady devastating cataclysmic period, maybe of just a few hundreds of years, and we will have been the instigators.

292. Turning the Corner:
Although it may seem unlikely at the moment, there is a new Australia around the corner, a new country that will show the rest of the world the way forward.
It is indivisible that one day soon we will turn the corner to live a life based on the Aboriginal doctrine. However, and even within the realm of possibility stymied by European conventions, we may be forced to another corner, and adapt the new Australian doctrine under some threat. If you want it to be beautiful, do not chuck shuttle at the picture.

The future can either be sweet or stink to high heaven, it's your blood choice, are you going to continue doing nothing and to be content in blaming others, or are you going to get of your hacienda and stop being a blood hypocrite.

Noteworthy:

293. Bull:
Where there are a many bulls, there will also be a lot of bull manure, and it is this that will clog up progress, to the extent that the clearance of this detrimental condition becomes rather difficult to perform, politically speaking.

By the way:

294. Warburton:

There are no trees, and probably no water, and Warburton is a waterhole in name only. The trees have been felled for domestic use, and the grass has been grazed out of existence, eaten by cattle long since gone, and they having left behind them a dust filled environment that will take a very long time to recover its former green glory.

Then the truth gates open, and we are given the chance to lean further toward the fantasy and the comfort it gives, with false attributes that go unrecognised. It would seem foolish to contradict the socially accepted norm of false beliefs and declarations, to go against them would amount to unrestrained social suicide, those that once were friends would become not so friendly, and then honesty would be allowed to melt away. However, the truth of being a real Australian could and hopefully will survive, and (re)surface as a new way to silence the untruths and misguided averseness of the strangers to Australia, the social 'incants' (delicately put) who deem it necessary to ostracise the truth at any cost, and they are really good at it.

295. Areyonga:

'Areyonga' – a tiny missionary settlement wedged between two sandstone mountain faces of the MacDonnell Ranges. The fish and chip shop only opens on the 29th of February.

Author's Extra Note: All the talk in this modern world of being good to each other is, I am really sorry to say; in some case, just that, talk. If we are all intent on being good to each other, listen to the rhetoric of an irate driver when the road in front is blocked by a cyclist, or an elderly person, and then think of being good to that driver.

Chapter 21. Nature Part 1:

296. The Everywhere birds:
Black cockatoos, sulphur-cresteds, swallows, Major Mitchells, willy-wagtails, quarrian, Kestrels, budgerigar flocks, bronze wings, finches, not to mention the odd galah or two.

297. Food:
'And there were kunga-berries, and various solanums and mulga apples and eucalyptus manna to eat as I walked along', and toilets everywhere.

298. Uluru:
Then I saw Uluru, I was thunderstruck. I could not believe that such a red form was real. It floated and mesmerised and shimmered, and looked too big.
The great monolithic rock was surrounded by fertile flats for a radius of half a mile, which, because of the added run-off water, were covered in lush green weed and wildflowers so thick you couldn't step between them. Then the dunes began, radiating away as far as the eye could see, orange fading into dusty blue. With the odd camper here and there.

Author's Extra Note:

'I sat on the first small mound of sand, watching daylight colours in luminous pastels, then a change with the blue and purple of peacock feathers.' Then I heard a peacock fart, and the smile on my face was genuine

299. Eucalyptus:
'Tallest tree in the world after the Californian Redwood'.
'How tall are they?'
'Up to three hundred feet; at an average of about two hundred.'

300. 'Kuzzie':
Kosciuszko 'Kuzzie' National Park in the Snowy Mountains, just over the border in New South Wales.

301. Australian Landscapes:
To be in Australia is to be amazed by space, and humbled by the most ancient, bony, awesome landscape on the face of the earth.

It is to discover the continent's mythological crucible, the great outback, the never-never, that decrepit desert land of infinite blue air and limitless power.

But anything would be mended, anything forgotten, any doubt withstood during a walk through timeless boulders, or down the dry glittering river-bed in the moonlight.

302. Kata Tjuta:

The rocks of Kata Tjuta are heavy dark and strong. They rise up like an island of natural power.

303. The Nullarbor Plain:

You can drive, walk, or in my case, 'scoot' through the trees into the huge Nullarbor Plain, and all around you will be an immense dry land with a 'population' of knee-high plants. Above you will be an even bigger sky, one that sits like a natural blanket over the land for as far as you can see, with your part in the landscape so minuscule.

At some point you will come across a diminutive spring, a tiny damp patch of the Nullarbor, where the Arangu aborigines sometimes gather to celebrate their ancestry with dance and song. Be respectful and be welcome, and the Nullarbor will not seem so unfriendly, perhaps.

304. Australian Soil:

The glory of the Australian interior is its soil. Its redness is of such ubiquity and intensity that is seems not only to stain the hands and the hair but, after a few days of exposure, to tint the whole world for weeks, as if the cerebral cortex has been injected with a solution of carmine red.

305. Claypans:

Claypans are temporary lakes, basins into which surface water will drain when there is sufficient precipitation, where evaporation exceeds precipitation; such basins are usually covered with water only briefly and shallowly for a few days each year.

What chiefly characterises claypans is their flatness, their surface being graded, and made fresh with each inundation of water. What if efforts were to be made to make the lakes permanent? Could be a good step in the right direction at least.

306. The floor of the pan:

We headed out of town through Darwin's sunny orderly suburbs - white bungalows on tidy lawns -
and at the edge of town passed a sign that said 'Alice Springs 1479 kilometres*. Ahead along the lonely Stuart Highway, lay nearly a thousand miles* of largely unrelieved emptiness all the way to Alice Springs. Good land for building new towns and villages, and parks and gardens.

* I ain't going to mess about converting miles to kilometres or the other way, live with it.

307. Daly Waters:

Daly Waters – 370 miles from Darwin, 570 miles from Alice Springs just off the Stuart Highway.

308. The Land:

c2018 'In the sunny middle distance, stood a large white farmhouse, surrounded by fields of unusually fertile crops.' Why unusually?

By the way:

309 Soil covered by dust:

Sheep farmers compounded the non-cultivating problem by increasing flock numbers, and compressing the earth with sheep-feet, for the land then to die from fierce overgrazing, and foot impressions by the trillions. This aspect of 'farming' left just a dust covering of the thinnest soil, and this blew away in the wind to leave the Australia we know today, full of rabbits and a bloke looking for a rabbit excrement free zone. Keep looking mate.

A trickle of a hint:

29 April 2019

310. Joseph Gerrard c1796:

'Whatever destiny awaits me, I am content. The cause which I have embraced has taken deep root, I feel, ultimately triumph. I have my reward. I see through the cheering vista of future events, the overthrow of tyranny, and the permanent

establishment of benevolence and peace. It is as silent as the lapse of time, but is certain and inevitable.
I will paraphrase that.

311. Destiny:

Whatever destiny awaits me, I am convinced the cause which I have embraced should take deep root, and help create a permanent establishment of benevolence and lasting global peace, There is a certain silence, an unspoken desire for certainty of peace and fulfilment of wellbeing and safety.

312. A waste of water:

Massive amounts of freshwater is wasted every second all over the planet. This water is provided through natural means and costs nothing, but we intelligent humans do nothing to utilise this life giving and life saving product, choosing instead to make use of oil, gas and coal, commodities that endanger life and the planet. To say we got it wrong is putting it mildly.

By the way:

20 April 2019

313. Melbourne:

Classy and busy in a cool way, and long may it stay that way; or has it changed?

21 April 2019

314. Royal Melbourne Institute:

Can I make contact with individuals who would be willing to make an effort to do what this world of ours needs? It could be that someone at the 'Institute' will be able to help me.

Chapter 22. Nature Part 2:

315. The Royal National Park:
A little further to the south, to the Royal National Park in the south, or the Blue Mountains to the west of the city, you will find raw landscapes that have remained utterly unaltered for thousands of years.

316. Christina Rossetti (1830-1894):
>Before green apples blush,
>Before green nuts embrown,
>Why, one day in the country
>Is worth a month in town.

317. Digging Ash. c1909:
It had been intimated that grazing and foraging farm animals had eaten the yam tubers out, plus the destruction with the soil becoming hardened with the tramping of sleep, cattle and horses. In that year the digging the soil was like digging ash.

318. Reddish Loam:
'The land was free of timber or scrub of any description, the soil a reddish loam of great depth'.

319. Murrong:

When Mitchell arrived at the Victoria Grampians in 1836, he saw 'a vast extent of open downs... quiet yellow with Murrong, and with 'natives spread over the field digging for roots'.

320. Captain John Hunter:

In 1789 Captain John Hunter, of the First Fleet reported that the people around Sydney were dependant on their yam gardens ... low banks appear to have been ploughed up, as if a vast herd of swine had been living on them.

321. The Service of Water:

Wentworth Falls and Sylvia Falls maintain a constant flow of freshwater to the lower areas.

The Hawkesbury River flows through green and beautiful lands just north of Sydney, with its sparkling fresh and natural waterways.

It must be said, water is the only commodity in Australia worthy of professional business acumen and forthright distribution, but it would seem there are far more important matters that override this.

27 April 2019

322. Countryside or Outback:

In many respects, the outback has not changed in over two hundred years, even though the white Europeans did their best to balls things up good and proper. Where there are the pre 1788 plants and creatures, they have had to be tough to survive, and in some cases, they have turned out to be downright dangerous.

With regard to folk, do not expect any special involvement, just good honest discussions and advice. By all means, 'pass the time of day', but most importantly take advice if it is offered, don't be a pillock, listen.

323. George Elliot (1819-1880):

Animals are such agreeable friends-they ask no questions, they pass no criticism.

324. The Flinders Ranges:

A range of hills on the edge of the Outback, over 300 mile (500km) from Crystal Brook in the south and Mount Hopeless in the north.

325. Karijini National Park: In Australia's desert heart, blue skies, red sands, orange rock and dry, golden spinifex grass, characterises Karijini National Park, a 630,000 ha (1.5 million acre) part of the Kimberley region. A permanent source of water can be found in the spectacular gorges in the north of the park.

326. Sheep, Camels and Rabbits:

There are millions of sheep in the outback and maybe a million camels, and the counting of rabbits if forbidden, as it is too depressing. All of these creatures are not native to Australia, and their presence has contributed to the country's demise in flora and fauna, in fact they are the main blood cause. Without them Australia would be a lot greener today than it is.

327. Flies:

The Australian flies seems to be of a separate generic stream of bloody mindedness and are a constant plague in Australia, their search for moisture is a search for life, and they will not shirk in this, they must not, they cannot.

Some of today's flies are descendants of those introduced with the sheep of two hundred or so years ago, but the black bush fly, a true native will never fail to make its presence known, in its hundreds and thousands. Take a leaf out of the book of the fly, and strive to survive. When we are gone, the flies will still be there.

328. Restoring the Balance:

Before the 18th century, the natives of Australia lived a balanced existence, and what is outback now was then hospitable and made friendly by judicious farming and land management.

To restore this balance, not just in Australia but worldwide, much must be done to revert back to a more natural lifestyle, not, I hasten to add, loincloth and sod huts etc, but by making the planet important again, along with animal, plant, and human life.

It is imperative that we humans adopt a more socially and worldly style of living, so to create a new spirit of involvement, one of being true to our own natural make up, with a productive and healthy respect for the environment.

329. Edible Fodder:

The edible fodder in some desert country is a commodity that is not there in over abundance, and when it's gone, it's well and truly gone, and some of it already has.

Over a period of not many days or even just a few hours, the desert harvest can be a pile of dead cattle, and that will not do anybody any good.

330. Scorched earth (Earth):

Scorched earth is everywhere in Australia, you can find it, and then you could watch it change to a magnificent land of a good, if still green-scarce, a panorama of true honesty, what you see is what you get my dear reader.

331. Aborigine life is free in the free country:
Physical, political and emotional aspects of living free as an Aborigine. Being careful without being holier than thou, an easy way to describe Aborigine life.

Aborigines were stripped of their souls, their elegance made to look wrong by white standards, standards that were flawed but in their turn made to look right. The whites then, in the 19th century of discovery, and even today in the 21st century, liken the Aborigine life to that of not much above animal life.

The Aborigine 'natural existence' is somehow in tune with their ancestors way of living, but with the white population regarding it to be an ignominious trait from which there is no escape, modern Aborigines have the uphill task. They are looking right into the whites of other eyes, while at the same time being true to their right to live as they please, just as other Australians do, but they have the extra burden of having to gratify other Australians.

332. Mungilli Claypan:
The sand is soft, and giant Ghost Gumtrees are seemingly idling about, and ignoring the heat.

The claypan is surrounded by low sand dunes, and a bordering rim of trees reflect the evening sun with a crimson glow, with their leaves shining with their greenness. The rose tinted glasses are returned to my pocket, and I move on, with the sand between my toes irritating me somewhat.

333. Choose your sand:

Who does that sand between your toes belong to? It ain't mine, and I am damn sure it ain't yours; even if it is, the next gust of wind could make it someone else's.

Chapter 23. Extra Future:

334. The motoring-future:
Travelling across Australia now, in a motorcar, is not too difficult if you take the necessary precautions. However, over the next 10 or 20 years, this comfort will have to be paid for, with Mother Nature setting the price, with heat and floods so catastrophic that the shiny new car you have just paid a ransom for will be about as much good as a plastic frying pan. A temperature of 30 degrees will seem cool when you are experiencing over 50 degrees, or even 60 degrees, with no trees to create the shade you need.

If they have any sense, and I do believe they have more than us when it comes down to it, the kangaroos will have emigrated to Antarctica, and the Emus will have learned to fly, what will you do apart from politicise yourself profusely, and sell your car to an Aborigine?

Author's Extra Note: It means nothing if you do nothing, but it will mean a lot if you do a little. A nonsensical but true historic account of what history has done for the Aborigines, and what social history as done for the motor car, two distinct opposites, where nothing and little are consequential.

335. The secret need:

The need for secrecy is very much one of the traits of humankind, and has been so since such 'creatures' began to think in an intellectual way.

They who are privy to a secret are in a more desirable position simply because of the fact there is a secret, but what is the secret for and of the future?

336. Do it for them:

Do it for the people we will never know. Like the Aborigine, we can prepare for the future of those who will come after us, in a pro rata socially driven survival pattern.

337. Deferred ending:

c2032. Just west of the small town of Bingara, in New South Wales, work was in progress on a new complex of dwellings based on an old Aborigine design. Old for new, to survive.

338. c2341 The Australian New Forest:

Stretching from Bingara in the east to Brewarrina in the west, and from Pembroke in the north to Mendoorah in the south, the New Forest was making life easy for New South Wales in a way no one could have foreseen. It was a good and life giving way that made global warming seem a distant and soon to be long forgotten nightmare.

339. c2355 The Simpson Forest.

Work is in progress on a new forest in the Simpson Desert of over two thousand square kilometres.

The progress is expected to be slow, but all involved, including each Australian State and Territory, are confident that, based on the Bingara model, the plan for this forest is sound, and the work will make the desert area a new and viable living zone for humans and animals within ten years, or less.

340. A classless society of two:

We should simply let it happen, an Australian classless society, a society that consists of those with money and property, and those with property and money.

However, in Sydney, there is the division of the harbour water - a dry and going on to hilly country to the north, and to the south is the low country.

To Sydney's busy people it would seem that whatever is to the north, or the south, has no bearing on society. Everyone is the same, choking with invisible deadly traffic fumes, and killing themselves with the worry of being one above the next person, a rat race with no rats, only those with, and those without.

Author's Extra Note: The cost of progress gets ever more expensive as society moves forward, and that same society seems to justify that cost, even if payment is in lives, which it is.

341. Dispossession or not:
Australia is a hard country, and it can be brutal and sudden in its treatment of those who disregard advice, but then, Australian's are no strangers to hardships.

It bodes well that changes are being made to soften any impact on Australia's struggling fraternity of Aborigines, and that they in turn are making inroads into compatibility.

This new partnership will be good for Australia, and good for the planet. With a new Australian approach to global warming based on thousands of years of Aborigine dealings with Mother Nature, there is the possibility of huge dividends, with the creation of a newer and more vibrant Australia, a country that leads the way forward into a cleaner and brighter future for all.

342. A bonus per head:
Idle contemplation will play no part in a new Australia; that kind of supposition belongs in the past.

What lies ahead is a more compatible nation, with benefits that are long overdue, and this is only what the Australian's deserve, and have always been within a gnat's whisker of achieving.

If the bloody wimbly politicians can get away from point scoring, and coalesce into a more sensible regime of a national and global dedication to do what is best for all, and not just those who can talk bullshit for hours on end, in an oration that means buggerall, we will move on to a better life.

The bonus is one of national and global reward, with each person receiving the benefit of a clean new life, in a clean new world, with Australia leading the way.

Chapter 24. Any other business:

343. Ano 1.'Jackaroo':
Australian maps 1:250,000 series topographical.

344. Ano 2. Aussie Zen:
'She'll be right mate', an Australian Zen statement. Everything will work out fine: if you get off your hacienda that is.

345. Ano 3. The freedom of 'Net thinking':
When the 'net' way of thinking became ordinary for me, I too became lost in 'the 'net', (Always remember a good idea is only good if you act on it), and the boundaries of myself stretched on for ever, I was free.

346. Ano 4. A Pause in Macksville:
Set on the bank of the swift and muddy Nambucca River, essentially just a pause on the highway to Brisbane from Sydney. Hold on a 'flung dung' minute, it might be just 'a pause' to some, but to others, particularly those who live in Macksville, it's a wee bit more special on a cool to warm summer evening.

347. Ano 5.Travel. (To Australia):
Could it be that the traveller was fearful of the new land, and stumbled in the search to get to know Australia? Maybe a new Mobility Scooter would help.

348. Ano 6. Menindee:
A modest hamlet on the Darling River; a couple of streets of sun-baked bungalows, a petrol station, two shops, the Burke and Wills Motel (named for a pair of nineteenth-century explorers who inevitably came a cropper in the unforgiving outback). There was also the semi-famous Maiden Hotel, where in 1860 Burke and Wills spent their last night in civilisation, before their unhappy fate in the barren void to the north. At least they had the imagination and the balls to try.

349. Ano 7. Hotel purchase:
One million seven hundred thousand A$ could buy a hotel in the outback. This is equal to £94,444 at an exchange rate of 1.80. [Transversely - £50,000 will buy a property worth 90,000 A$.]

350. Ano 8. Counting Aborigines:
In a 1976 referendum, Australians voted to change their country's constitution. As a result, for the first time, all Indigenous Australians were also counted in the federal census,

and came under the legislative protection of the Commonwealth Government.

351. Ano 9. Mabo:

In 1993, after the Mabo decision of 1992 by the High Court of Australia, the possibility of native title was formally recognised by the federal Parliament.

352. Ano 10. Charles Sturt:

'Everything tends, I believe, to prove that a large body of water exists in the interior'. He would have been right, if only he had known at the time.

  Charles Sturt.
Still looking.

353. Ano 11. Population at peace:

Australia is mostly empty, and a long way from most places. Its population is small, but it is a stable, peaceful and good place to be, even if you are an Aborigine living in hope.

354. Ano 12. Distance:

Australia is an interesting place, with thousands of miles or kilometres between places of interest, and between places of not so much interest, I suppose the space between can be more compelling.

355. Ano 13. The 60,000 year-old Aborigines:
More investigations has made the aborigine's time in Australia to be in the region of 50,000 years, maybe as much as 60,000 years, with a hint that it might even be 80,000 years.

356. Ano 14. Biologically speaking:
Early explorer's observation.
'Australia was deemed biologically deficient, its semi-arid plains too monotonous, and its forests too silent'. With its politicians too single minded, and that's me being polite.

357. Ano 15. Once upon a time:
For sand to be there, even if it is red sand of no nutritional value, there must have been a good top soil once upon a time. If good top soil can be changed to sand, why not reverse the process, or are there too many political or scholastic twits in the way.

358. Ano 16. A nation of folk:
You live in a nation of folk who are different within the nation, as individuals, consequentially; the nation is not the individual.

359. Ano 17. Bingara Gold:
The discovery of gold in 1852 brought prospectors to the area. In the 1880s, copper and diamonds were also discovered, causing a rapid development in the town.

360. Ano 18. Reading of Books:
Someone told me that Australian's do not read books. I find this difficult to understand, I would have thought they had needs for libraries just as much as we do in England. Maybe I will put it to the test one day with the editor of the Bingara Advocate; send her a couple of my books to start a library, and wait for a reaction.

361. Ano 19. Jane Austen:
To sit in the shade on a fine day and look upon verdure is the most perfect refreshment.
 Jane Austen (1775-1817)

362. Ano 20. The Sacred Uluru:
Nothing should be removed from Uluru, including stones or sandy soil. According to legend, the belief of the Anangu, everything needs to be where it was created, or else the spirits will not be able to settle. Just stand look and listen. You don't have to be a would be spirit guru to understand, just accept what your are told.

363. Ano 21. Watering Hole:
The original Alice Springs was a watering hole on the River Todd, the mostly dry River Todd.

364. Ano 22. Faffing about:

With satellite navigation, you will always know where you are, and with modern transport, you should be able to avoid trouble with consummate ease. Just as long as you don't faff about.

.

365. Ano 23. The Northern Territory. (NT):

The NT covers about one sixth of the Australian Continent, with an area of 1.35 million sq km, which is equal to the combined areas of France, Spain and Italy.

366. Ano 24. The Macdonell Ranges:

In central Australia, the NT contains the east-west ridges of the Macdonnell Ranges, with 'hills' that reach heights of more than 600 m. The well-known monolith, Uluru (Ayers Rock) is 348 m high and is near the south-west corner of the N.T. A long hop and a big skip from Alice Springs.

367. Ano 25. Happiness:

Is it possible that being happy and being 'Australian happy' still applies?
Happy: Enjoying or expressing pleasure and contentment (Description of an Australian?)

.

368. Ano 26. Photograph:
Harold Cazneaux.
1878-1953

'Children of the Rocks – Argyle Street' and Australian National Gallery Canberra.

369. Ano 27. Painting:
Albert Namatjira
1902-1959
Ghost Gum, Central Australia c1950
Art Gallery of South Australia in Adelaide

370. Ano 28. The Royal Society of South Australia:
Can I communicate with such dignified folk and make a difference? Can they communicate with each other and make a difference? If they communicated with you would you listen and take heed?

371. Ano 29. The Horizon.
If there is a blue shimmering on the horizon, all around you, then it is probable that you are in the middle of nowhere, and this can only change if you move, in any direction maybe, but try to make it the right one. Don't forget, if the sun is in your eyes in the afternoon as you travel, it will burn your arse in the morning. Therefore, with a following wind, and with sore eyes and a hot beam end you might make it to where you should be.

372. Ano 30 Parliamentary rule:
Any ruling parliament is just a puff of misrule. There is a vacuum in Australia, and it is only right and just that the Aborigines of Australia should help fill it.

373. Ano 31. Energy:
Conserving energy in a desert situation seems an obvious dictate, but there is a hidden significance, until:

Author's Extra Note: Because I am disabled, it does not mean I can't take steps, I assure you, I can, and what you are reading is proof of that. My hope is that they are steps that will help to save this sweet Planet of ours. If we could all just take one single step in the right direction... wow!

374. Ano 32. Taking steps:
'My feet were a part of me but they seemed to be in another mini-world of their own.
'Every step I made was made without full or partial deliberation, my energy level was so low that my power of thought diminished to such a state that my feet did what they needed to do; but in no way did my knees agree with them'.
But I managed; I just had to. It was a hard lesson, but it was a valuable one'.

375. Ano 33. The European or non-Australian Way:
The European society is chock bang full of bigotry; most of Australians have known this for a very long time; knowing that their European counterparts are of a different social type, with paranoia paying heed to intolerance of other religions and ways of living. However, and this is probably the crux of it all, this bigotry is still festering away in an otherwise liberated Australian social order.

Further to this is the continuing adherence to an 18th century unconstitutional philosophy that instigated full intolerance of non European peoples, and in doing so isolated and exorcised the Aborigine population of the Australian continent without exception.

Noteworthy:

376. Good wording:
As in literature, where a grammatical error will overshadow the genius of a new writer, so the bullshuttle smothers the foresight of those with honest goodness and worthy dedication and intentions. Weave this into the political spectrum and you have a mixture of fertilisers that are of no use whatsoever.

Chapter 25. Aborigine Part 3:

377. Resources:
Sustainable use of resources: No amount of European farming no-how can match the nuance of the Aborigine's method of farming.
In the Aborigine method there are no machines, just the knowledge of the land and a partnership with it.

378. Possibly Aborigine:
To have a friend you can trust is a real blessing. To have a group of friends you can trust is even a bigger blessing: However, if you live in a community of folk you can trust, you are possibly Aborigine.

379. Farming old style:
With the use of fire, the Aborigines cultivated certain areas of land to produce grasses, and other areas to produce root crops. They created grassy areas close to their settlement, making them suitable for the kangaroo and other animals, in this way keeping the 'meat market' within easy reach. It's not rocket science is it. The non-Aborigine way is to throw money at the land and let Mother Nature chuckle at such feeble efforts.

380. Missing tricks:

In the Aborigine world, missing a trick would be a matter of life or death. So maybe they get rid of tricks, and therefore they have no tricks to miss, easy isn't it.

381. Ancestral law:

The genius of Ancestral law was that people of a wide region, like the inland areas of Australia, could agree to a body of rightful edicts, without there having to be legislation. This was in spite of the autonomy of individuals and family groups, and certain ritual practices, and this tended to induce in young people a disposition to conform to shared values and norms, and to respectfully refer to Elders for arbitration, if it was ever needed that is.

382 No claim to any land:

Aborigines could build homes, sew clothing, build fish traps, cultivate root and grain crops, manage livestock, and live a peaceful coexistence with other clans, without the need to build fences, and this was thousands of years before 1788.

Some of the skirmishes that came about after 1788 did so because of enforced actions made by the invading Europeans. These folk assumed that because there were no fences on the vast land of Australia, and the Aborigines were nomads with no claim to any area of land. A policy of 'what's mine is my own, and what's yours is mine' seemed to prevail, without any problem to the new Australians.

383. Aborigine Elders:
To be an Elder is not an electoral procedure. It is rather to be a worthy one, known to be fair, judicious, and well versed in tradition. In a society that values these attributes above others, that is near perfectas we can get in a true democratic sense. In the Aboriginal society, an Elder becomes so by virtue of a long process of initiations and example, by earning the respect of others, and not just of their own clan

Author's Extra Note: This type of self-governing protecting of people and the land, in fact, all that appertains to the peace of existence over a vast area, and it is one that lasted over thousands of years, and if left alone by the all-knowing white folk, it could continue to protect.

384. Aborigine Peace:
Making cultural and dynastic changes without resorting to conflict, with no violence and no displacement, is a correct pathway to peace, and the Aborigines have been on that path for 80,000 years. Then the European folk arrived with their holier than thou attitude, inflicting foreign idealistic muck on perfectly happy people, people who had no understanding of these foreigners, with their proliferation of 'do unto others as

you would unto yourself as long as you don't do it to me,' attitude. If it wasn't so blatantly bigoted, I would say the bloody 'Poms' of that era were under the influence of idolatry dishwater, and drinking it in huge quantities. Do they still sell it in Sydney?

385. Aborigine abiding before 1788:
There were no wars of invasion to seize territory until the Europeans arrived.
There were intertribal battles or skirmishes.
They did not enslave each other.
There were no master-servant associations.
There were no looking-down-the-nose class divisions.
There was no property inequality.
There was no income inequality.

386. Thomas Mitchell. An interpretation of his writing of the Aborigines after 1788:
'These unfortunate creatures could no longer enjoy their solitary freedom; for the dominion of what folk surrounded them. They were hemmed in, deprived by the power of the white population, and suffering at the loss of their liberty. They formerly enjoyed wandering at will through their native wilds, but were compelled to seek precarious shelter among the close thickets, dry river beds, and rocky vastnesses which afforded them a temporary home.'

387. A Rock:
You cannot say a rock is just a rock in the Dreamtime Zone. You can only say, 'there is a sitting rock', 'it is a rock that leans', 'it is a rock that as fallen over', or 'it is a rock that is lying down,' but never just a rock.

388. Aborigine Empathy:
'The Aborigines seemed to be in perfect empathy with themselves and their country.' Then the convicts and settlers came along.

Author's Extra Note: It is maybe not right for me to comment on an event I have only read about, but the thought of men, women, and children, used as 'mules, then disposed of by putting a rope around their necks at Myall Creek. They were then dragged to their grave.

389. Myall Creek:
Myall Memorial Site
Whitlow Road
off Delungra Road
near Bingara
NSW 2404

390. Administrations and the Aborigine:

How is it done without an administrator? Well, there is less of the bloody paperwork for one thing, and no digital make-amends by being good on a keyboard, or some other electronic aid.

An Aborigine Elder did not need either paper, or a blasted keyboard. He or she relied solely on commonsense, and their peoples' history, 80,000 years before the computer was even thought of.

391. Dance:

A story in a dance, and a story from maybe many thousands of years ago, or even more thousands.

To comprehend this storytelling we must first disregard all matters modern, which are subjected to unfair, and in some cases downright bloody obnoxious doctrines, having been brought on by scholastic knowledge that bears no resemblance to real life.

392. Respect:

'Intelligent harvesting'. 'Conservation on a vast scale'. 'Sustainable social ability at very little or no cost'. These three attributes of the Aborigines are due much respect from the rest of us. Not to show the respect is tantamount to downright ignorant behaviour that is very close to lack of common sense, or, if you like, bloody stupid.

393. Management:
For tens of thousands of years, the Aborigines were the managers of Australia, with 300 or so different tribes or clans unified in the task of keeping the country in good order.

394. Legitimacy:
If we assume that the Aborigines were, and are the legitimate original Australians, does that mean that the European convicts, settlers, and later immigrants were and are part of a suppressive society? If so, could there be a certain level of criminality involved in keeping down the legitimacy of the Aborigines. This can easily be resolved: get rid of all kinds of racially prejudiced and historic belief that the religious dogmatic actions of the late 18th and early 17th centuries were based on Christian ethics of goodness, they were definitely not, they were more holier than thou and using unadulterated bullshuttle.

395. The ABC of the 19th and 20th century Australia:
(a) Untold numbers had made it to Australia and then become Australian, with no connection to anywhere else in the world.
(b) A further undisclosed number of Aborigine Australians were discounted has being Australians, with no connection to anywhere else in the world.
(c) The distinction is obvious to anyone with a sense of justice, but it would seem that even-handedness had taken a holiday.

By the way:

396. The Early European Convict/Settlers:
They were launched into a world where speculation reined supreme. They were given an opportunity, but it was one they were not equipped to recognise. The warm sunny weather of the sunny south was overpowering, and any down to earth mundane considerations were swept aside by a new and worrying sense of survival. As each day went by they determined their future by exercising a cruelty they would have hitherto known only in the isolation of an English, Scottish, Welsh or Irish slum, where condition were far below the accepted norm of their high minded superiors. It was as if the were intoxicated by so much freedom in such an idyllic arrangement, and by making the Aborigine of less worth, they made themselves more worthy.

Noteworthy:

397. Demise or magnificence:
It is unthinkable that the populace of this planet will perish, but in fact, this demise is already in progress, and has been for some time, with super dry countries and continents leading the way.

Life in the middle of Australia is becoming extinct, with the soil unable to give root systems a foundation for growth, making it a signpost to the future.

For perhaps thousands of years, the area was holding its own. However, the climate has begun to change, driven by human action and non-action, a modem of interaction of greed and so called modernisation.

398. An essential part of the world:

The far distant horizon tells me I am in Australia, and, of course, I am still in an integral part of the world. However, there is the fact that my disability and my Mobility Scooter could be a wee bit of a disadvantage, and the writer I am may not be bullish enough, but at least I have had a go.

Leslie H. Harvey of Kimberley.

Chapter 26. Notes on New Australia

NEW 399. Arrogance – The Australians:
By being calm and being arrogant, the new Australians make the new world a fresh and dedicated unconfused future for all of us. They lead the way, and all we need do is follow, keeping our minds on what the future can be and will be.
Calm arrogance is steady self-awareness with assurance, and in this way, the world moves forward confidently, thanks to Australia.

NEW 400. The dumping grounds:
What had once been the dumping grounds of Australia, the dry 'couldn't give a bloody fig about' areas of desolation and desert, are now becoming a hinterland of wealth in sustainability.
Where dry creeks had failed to retain a constant water supply, changes to water flow have made them viable irrigation tools of a huge environmental importance.
Where once the Aborigines wandered and camped in desolation, there are green and floral areas of natural integrity.

NEW 401. Clacton Wood:

In the year 2026, the place, Alice Springs, just over 200 years after Billy and Sandra Clacton had begun the family and the farm, and the legend they became, Stan Clacton, their descendent, walked over from the home farm to woodland he had known from birth.

Clacton Wood was a scene of green bushes and trees, and birds with their colours and singing and chatting, with a small lake completing the cool ambience perfectly, a true legacy of the legends that were Billy and Sandra, and the essence of New Australia.

NEW 402. The Inland Sea of New Australia:

At the beginning of the 21st century, the promulgation of just a wee bit of water in the dead centre of New Australia was just a forlorn hope, and not even anywhere near a dream.

Then, using the return river water system; pumps driven by windmills taking freshwater from near the mouth of a river, and channelling it to the parched middle of the continent, the water supply was seen to be continual and uncontaminated except for a high nutritional value.

As the water level grew, the hitherto infertile land began to show signs of new green and high dietary growth, on a scale not thought possible.

With each passing year of this return river water reaching the New Australia centre, a sea, or more correctly, a lake, began to form, one of a huge size and depth, with the land sloping down to the water supporting grasses and trees in an ever-expanding profusion.

NEW 403. The Quiet Trees of Bingara:

'To sit in the shade on a fine day, and look upon greenness is the most perfect refreshment'.

Jane Austen (1775-1817)

She would have written those lines around the time of 'White' Australia's beginning, and they are just as true in the middle of the 21st century and in New Australia.

The Bingara Woodland, a huge expanse of trees that give shade and sustenance, is where the Myall Creek is tranquil and quiet.

It may not be the same woodland that Jane Austen spoke of, but it is made up of trees, trees of all shapes and sizes, Australian trees that provide shade and sustenance for those who seek the same shade and greenness she experienced – life goes on in peace, so it would seem.

Chapter 27. Nation and Lifestyle:

NEW 404. Honesty – The New Land of Australia:

The new Australia is in many ways preferable to the old, to the unrelieved past of doubt as to the possibility of a future that had prevailed for more than two hundred years. Of course, there had been times when Australia stood proud, but at a fearful cost of blood and tears. Now the new Australia can be proud of a peaceful expense of making life good for every Australian.

NEW 405. Changing Aborigines:

The omni presence of Aborigines who could see no wrong in taking what they needed cannot be denied, their lives had been governed by this principle for 80,000 or so years, and changing was not just difficult it was anti social. Nevertheless, change they did.

NEW 406. The business of living:

Living to work – or working to live. The latter was a statement of truth for many folk prior to the 21st century, but now as we move further in to the new century, Australians are doing neither, instead they are living to be happy and content, in an environment of civilised dealings.

This is not someone's dream, it is a reality, and if you are a new Australian, then why not?

What can you have against enjoying life by working to a new timetable of ease and full cooperation?

seek not work for work's sake, but work to seek fulfilment of joy for a better life made possible by a wiser and better security.

NEW 407. I was no privileged person:

I have never been privileged to an easy life, my only recourse to care for my wife and family was to work, and work hard, sometimes up to 18 hours per day.

Because of my struggling, I can understand, and comprehend with sympathy the plight of the early Australian, black and white.

My knowledge and experience of hardship, and I can write and hopefully portray a hint of the Australia that will now lead the world, honestly.

NEW 408. Ball and Chain:

Acquiring wealth is perhaps a worthy endeavour, but if by doing so the task becomes irksome and overpowers common sense, then that worthiness is replaced with a time consuming and heavy responsibility, time being the ball and responsibility the chain. Life becomes less rewarding in enjoyment, and more dogmatic in operation.

The worth of one can mean the loss of another, with never a balance, but weirdly the loss of both can make for you and yours a happy contented life.

NEW 409. Keeping your nerve:

The opposite of being wrong is to be right; there can be no other opposite. If you know you are right, keep your nerve and remain correct, to defer would be an error, and that would make you wrong, and that will not do. It's not bloody rocket science, as they say.

The success of New Australia depends on keeping to the schedule of being right, a schedule that lifts Australia above the rest of the world, the 'rest' of the world who have got it bloody wrong for whatever reason; usually the wrong reason, and got it wrong big-time.

...

Author's Extra Note: Living in a country that can claim to have a utopian society is probably the main reason Australians can be proud, so don't let the wobblers pull you down.

NEW 410. Fear, anger, and worry:

Where there was once a few, there are now many, and where there was once fear, anger, and worry, there is now a new underlying excitement of beauty, and the New Australia of sweetness is active and doing really well thank you.

NEW 411. Clacton Farm:
The year, 2026, the place, Alice Springs and the Todd River.

The Clacton family had their roots in 'Alice', roots that went way back to 1822, when Billy and Sandra Clacton, fresh from England, via Melbourne, staked a claim and made their home at Clacton Farm, at was then 55 miles to the west of Alice Springs – it is now so many kilometres.

For year after year, Billy and Sandra struggled to make the farm pay. Some years it was okay, but others were nigh on impossible, with heat and dust the only 'crop' for months on end.

NEW 412. Beautiful country, beautiful people:
Living in New Australia is one life long holiday. There is work to be done, but this is shared among the fit and healthy, while the disabled and old folk are cared for and understood – no one is a loser, all are winners.

Of course, the bigger the effort the larger the reward, and some folk, while staying loyal to the New Australian concept, make their own lives that much better, and has time moves on these more fortunate folk have become the norm by which New Australia can be justly proud. No one is left behind, and with each step forward, a helping hand makes life not just better but fairer.

NEW 413. Resistance:
To resist is to try to put an end to what you don't like. But if this be the case, could it just be a personal thing, and have nothing to do with the overall picture? Are you being selfish? Is it your way, or no way at all? Then, it could be just a load of bollocks, and the country is too big to be bothered about your hang-ups.

The resistance is turned on its head and you are the one who is out in the cold and being of no use to anyone, including yourself.

By all means have your say, but be big enough to know when, to resist is to balls things up for everyone else, get with it and go with the flow.

You could stamp your feet, or say 'I want', but your resistance will be no more than a pebble drop in a huge smooth lake of compatibility.

NEW 414. Different or not different:
A peaceful early morning in Kimberley, England, in July.
The sunrise is perfect in a blue sky, with not the slightest breeze to disturb the ferns and flowers of the woody green valley.
A peaceful early morning in Bingara, Australia, in January.
The sunrise is perfect in a blue sky, with not the slightest breeze to disturb the ferns and flowers of the woody green valley.

There are 10,000 'distances' between the two mornings, and that is the only difference. We could assume that this is the epitome of a world in harmony with Mother Nature, let us hope the 'mother' agrees with us.

NEW 415. Seldom seen or heard:
As in every village, town, city and nation across the world, there are, around New Australia, good honest folk who would do no harm to anyone, and, in some cases, to any living thing. They are probably in their millions and live in every corner of the country, keeping quiet and staying low, and out of mind. It beggars those who mean well, and are noisy, and can be seen in most places, even in te lower of establishments. These very visible and noisy people are, by the very virtue of being seen and heard, giving the wrong picture and sound of New Australia.

NEW 416. Proud to be fair-minded in New Australia:
Contrary to some folk I know, I don't give a stuff about 'pride coming before a fall', and suchlike gibberish, but I do care for folk who say they are 'fair minded' and behave so.
If you spotted somebody struggling, and not making any headway, would you think 'tough shit mate', or would you see if you could help. Whatever you do, you could be chastised. First, for standing by and doing nowt. Second, for poking your nose in where it's not wanted.

It's not a dilemma as such, it's just human nature. Do what you think is right, and don't worry about the pillock brains, just be proud to be fair, and remember, when the manure is about to hit the fan, duck.

NEW 417. Standing on Collins Street in Melbourne:
The movement of people is like the easy flow of a shallow stream, with life on course for each one there.

Yes, there is noise, city noise, sounds of people talking, and the trams and traffic moving from wherever 'A' was to wherever 'B' is. Thousands of journeys, none of them straight or without pause.

NEW 418. The enigma of new versus old:
All through the history of humankind, there has been the enigma of youthful innovation as apposed to the maturity of the old. Whereas the young can pave the way forward with innovative thoughts and actions, the elderly can bring a sense of having seen it all before, and have significant details of what has been tried before, and what was successful before.
Unite the two ideals, young and old together, with a harmony of people-power as apposed to power for its own sake. With both acting as 'Elders', common sense will rise up from the ashes of materialistic delusions of misguided grandeur.

Being young and able-bodied is a blessing not to be ignored, and being mature is a privilege that all should recognise and benefit from, with no exceptions; racial discrimination has no part in the future of humankind.

NEW 419. Compose the future: A colony of ants are composed, they gel into one living organism with thousands of individuals, but even though each one is free to move and do things as an individual, in isolation, it is the welfare of the colony that is paramount.
These tiny creatures could teach humans a valuable lesson, one that we should be glad of ,as it is one that will save all human life, and the planet we call home.

Instead of pulling every which way against each other, we need to compose ourselves and work effortlessly, without any form of threat or persuasion, towards the one true conclusion, an end to the threat of global warming, and bring in rye future with the melodies of a natural living harmony.

Author's Extra Note: New Australia is in a really good position, head of the planet's 'civilisations', and with a gentle swerve to make this peace and prosperity happen, New Australians will show the way without fuss or bother of any kind, just easy moving changing lifestyles across all of New Australia.

NEW 420. City limits:

Right up until, the beginning of the 21st century, every city in Australia had its peripheral limits, either hemmed in by sea, or cornered by the hinterland that had few friends. However, with New Australia the restriction of a closeting land mass has been controlled, and the grass and trees that have replaced the scrub and bush land now make it almost imperative that the city limits should be extended, and the countryside around made settlement friendly. As a result of this new approach, new towns and villages are springing up, with new wide tarmac roads, and building plots aplenty for an expansion into the hinterland.

In line with the expansion, controlled rivers will be supplying all year round freshwater, canals are being built for irrigation, and in amongst the towns and villages there are new arable farms growing a plentiful food supply.

Chapter 28. Water:

NEW 421. Locked Water Levellers:
With the use of locks, 'water levellers', the fresh water can be lifted and flushed clean at any stage of its transition across the country. Fresh water in the rivers is diverted before the water becomes saline, via canals and locks, with the use of windmill driven pumping stations.

As the system is expanded, the Australian outback has become greener and more productively greener with grass and trees, and productive in vegetables, fruit, and grain, a veritable utopia of farming country not seen in Australia for hundreds of years.

NEW 422. The constant presence of canals and rivers:
Where there was once dry arid plains and valleys there are now freshwater canals and rivers. their presence is making the countryside friendlier and, of course, more productive in all kinds of ways.

The tourist is pleasantly surprised to find cool areas of woodland, and scores of open green fields full of friendly vegetation.

The canals offer every chance to see Australia while staying afloat, and living in cool comfort, on a journey that less than twenty years ago would have seemed foolish to dream of.

NEW 423. Exchange is no robbery:

The freshwater of the creek makes its way down stream to the river in huge amounts during a storm, and good clean water rushes down river to the sea or to overspill into the low lying floodplains to cause flooding havoc. Then the storm abates, and river level falls, eventually to leave a dry creek. However, in New Australia an exchange has been made; whereas the water flowed headlong to the sea, windmill driven pumps now move it to canals, where the water is regulated by locks and weirs, to return inland to where it is most needed.

Chapter 29. The Political Miasma:

NEW 424. Coalition, non-political, and non-academic:
The coalition style of leadership is prevalent at every level, in cities, towns, and hamlets throughout Australia. The feudal nonsensical back biting and squabbling has gone, and the word 'political' and its connotations have been banished.

A new handshaking regime is prevalent at all meetings, and agendas are stabilised to cut out time wasting arguments.

NEW 425. New Australia without 'democratic' interference:
Careful application of common sense is not out of the question, when political folk are making inroads to a sensible of system of leadership by reputation and not by definition. However, with a non-political leader, an outright sensible 'elder', a determined choice will take optimum control by a non-discriminatory policy indication.

New Australia is now in a utopia of elders who make life easy by reputation, and in this way all who are part of the social living are gaining in justice and equality.

NEW 426. Political fear and retribution:
The act of being wrong without the risk of retribution is a political trait. With that said, there is the political courage to instigate changes, hopefully changes for the better without relying on 'big money' cracking the whip of compliance, although the one usually goes with the other.

Then came the first summer of real change, the first summer of New Australia. Not only were doors and windows opened to a fresh new warm and sunny morning, the whole country opened its eyes, and turned its back on big money political catchalls. Politics were dead in the water, no one wanted to know of promises, not when the new Elder system guaranteed social equality and an Australia with a single mind, with the thought of prosperity and a healthy environment for all paramount.

NEW 427. Political identity:
Of all who call themselves New Australian, politically or otherwise, the non-political Elders are the true consciousness of national awareness. Conversely, this also acutely noticeable among the young spirited New Australians.

The national flags are not put away, they will crease if folded, they are flown proudly, everyday and night, and can be replaced with the bright New Australia flags when required.

NEW 428. New Australian bywords:

There is some things that do not exist in any other parts of the world, for example peace and plenty. These two are the bywords of New Australia, and no matter what the rest of the world may think, say, or do, these bywords will ring true in New Australia for a very long time.

Whereas some countries manage to live in a kind peace, they miss out on the plenty. Political interference in local commerce, and attempts to do likewise on a national basis, have resulted in legislation that feeds predictable bull manure to the civil service in huge clumsy amounts, and while the political civil service prospers, the people they are supposed to serve and keep from harm are struggling just to stay alive and well.

The old Australia was of one of the countries missing out, until the Elder system of social awareness broke through the long distances of bloody ignorant red tape, it was then that New Australia broke free from the constraints of over egged bureaucracy. The new social order became more egalitarian than anyone had ever thought possible, an occurrence not experienced since the Aborigines had worked with Mother Nature in the thousands of years of their pre European Australia.

NEW 429. The opposite of front:

Of course, 'back' is the opposite of 'front', but a genius of a politician would make it much more clear.

Let us be clear about this. We of the Balderdash Aggravation Party are making it our policy, as detailed in our new manifesto, that by an internal definition to be in at the front is to be clearly of a more forward looking amalgamation of the truly democratically elected.

It should be remembered; the opposition party have a definite inclination to be at the back. Their manifesto, such as it is, cannot be clearer in defining their backward policy of being all for the party at the front, a party consisting of a group of individuals not inclined to share.
We of the B.A.P. will stand by our manifesto as we take our seats in the senate.

Author's Extra Note:
In New Australia, such bull manure will fall by the wayside, and wither away in the gutter of discontent by moving forward to get to the back, or by going back to get to the front. In other words, it's all a load of unwanted politics, and politics play no part in New Australian life.

Chapter 30. The Global Effect:

NEW 430. Keeping the Planet safe:
Without being blasé about it, Australia is not just leading the way; it is playing a huge part in keeping the planet safe. The country's new greenness is making global warming a thing of the past, a devil of a threat to our existence has been removed by the people of Australia making the first moves, and continuing to make new and more global friendly decisions. Where 'global warming' once took the highpoint in discussions, 'global friendly' is now paramount.

NEW 431. The true worth of History:
If we are meant to learn from things that have happened before, why then should we concern ourselves with non-events. The non-event of global warming did rank high in the annuls of what was to become our recent history, but it was a non-event, it did not happen, and Australia is proving that history can be made to work for us in a non-active way.

NEW 432. The Thatched Cottage:
It would have been thought impossible that a thatched cottage would appear in the bush country near to Bingara, or in the bush country anywhere in Australia.

Like a gentle introduction to the new Australia, the white painted cottage on the Delungra Road out of Bingara sits quietly, with its thatched roof mixing in with the landscape. The local folk enjoy looking at the cottage, and the owners welcome enquiries from prospective cottage builders, with pleasant thoughts of more such cottages springing up in the Australian countryside.

NEW 433. A biased opinion:
Rid Australia of the unfounded bias that has been allowed against others considered o be unworthy of being Australian, and take what has been offered on a plate for the last fifty years or so, in a new magnanimous Australia.

There is no thumbs up rear thinking, not when the arid and barren landscape begins to turn green and floral in a never ending wind and rain of permanent change.

NEW 434. Dust! What dust?
It's gone. The dry bloody eye blinding dusty earth has gone, and the grass is the honest type that signals the loamy soil of success.

Place the palm of a hand down on true-cropped grass, and feel the life coming from the moist loamy soil beneath. That is the New Australia, green and as fresh as the dew of every new Australian morning.

NEW 435. A new approach:

When it was thought, and often verified by unwitting scribes, that the outback was a weird and unfathomable wilderness, hardly anyone argued the point. However, the global warming saga has begun to retreat into just a memory, and what was once the rough unproductive outback is now changing to green hills and vales, with woodland and forests intermingled in the fresh New Australia hinterland.

The new approach has worked, and even better than anyone could have hoped.

NEW 436. Smooth amidst the rough:

Hard by a clutch of rocks, some of them giants, there is a 'desert' of smoothness, a green expanse that goes on as far as the eye can see.

Close to the rocks, the soil had reformed and was now supporting the green growth above it, a change in soil from sand to a good loam had made the difference. Some areas are greener than the rest showing that the soil was even richer in loam and nutrients in certain places, and these areas were getting bigger with each season, with some already measured in square kilometres or thousands of hectares, and as they expanded they were supporting new tree growth with every passing season

NEW 437. Count in millions:
Obviously, a million good folk are better than just one, commercially speaking, but when those million are spread across New Australia, they count for much more.

Their presence means a lot to the new nation, with their numbers serving to signify how the joy of a good life can make a country bloom, as it moves ever onward in true harmony with Mother Nature and our Planet Home. The benefit to the people and the nation can be counted in millions.

NEW 438. Added incentive:
It is the year 2021, and a small child in a village Africa is close to death. The mother looks on helplessly as her child struggles to breathe, watching as the little one feebly reaches out with a withered arm and hand for the help the mother knows she cannot give.
Thousands of kilometres away on the continent of Australia, and unbeknown to the mother, a group of Australian people had made the decision weeks ago to send help to that village in Africa, barrels and barrels of fresh clean water, and boxes and boxes of nourishing food.
With perhaps only a few hours to spare, the water and food arrive in the village, and in the days that follow, the mother sees her child slowly recover, a smile returning to its face, where it had been too malnourished to even cry.

NEW 439. Centre of an interstellar storm:

It is the year 2019AD, and whether we recognise it or not, we are in the centre of a spiralling interstellar storm of a size beyond our reckoning, one that can be so destructive we cannot even begin to understand. Much better not to go there at all, but back off in fact, slam into reverse now.

The destruction this human storm will do to our planet is far beyond anything we could judge it by, far beyond our imagination, or even computerised imagery. Whatever scientists say it is just guess work.

The first indication of the destructive nature of this storm has been experienced in Australia over many years, but not recognised for what it really is. It is a blood chilling warning we should acknowledge and do something about.

However, New Australia and its people can and will show the way, and lead the world to the sanctuary of common sense, with the eye of the storm widened so as to make its potentially devastating tendency null and void.

NEW 440. Are you one who lags behind?

A dawdler can be defined as one who hangs around, hangs back, for whatever reason.

If, by lagging behind, you avoid a problem, it could be argued that you are using a kind of self-preserving logic – logic my arse – you are being a total fart of an idiot of the first order.

How would it be if we all adopt the political equivalent of lagging behind the 'let's be clear about this' defining pause-button pressing, a dexterity all politicians have a tendency to resort to when confronted by a difficult situation.

Get of your non-political backside, get the New Australian message, and catch up with working alongside Mother Nature and not against her. You don't have to be a politician or an academic to save our planet, just get-with-it, get with the New Australian way.

**NEW 441. Thin surface layers and a key:
New Bingara NSW**

Over the hinterland of New Australia is a layer of 'dead' material of a varying thickness, and it bodes well that where that layer is at its thinnest, a new covering of good deep loam is changing the properties of the land, and making it so it can produce plenty of food and fodder.

In line with the hypothesis that the loam is of a thickness to promote a new huge expanse of viable farming land, then it must follow that 90% of New Australia is ready for an agricultural expansion never before experienced anywhere on the planet.

Beneath any skin, whether it be human, animal, or earthly, there lies a dormant miracle waiting to happen.

The people of New Bingara are using the 'key' and have opened the door to a new and good future by joining the natural drive to save Australia and the planet.

NEW 442. Intimidation:
Sitting, watching, enjoying a hot coffee, just minding your own business and up comes a pillock with a mouth, a mouth he finds difficult to control.
The constant waft of words begin to grind on you, and you want to clout the sod to shut him up; instead, you smile, drink as much of your hot coffee as you can in one last gulp, and then you walk away.

You have just been intimidated by a pillock, one with a loose mouth and not much brain. Not a pleasant thought.
A thousand kilometres away, in the central desert region of New Australia, what was once arid and unpromising land as been intimidated into being green and promising. Now, that is a lot more of a pleasant thought, especially when you picture the fruit trees growing there. Welcome to New South Australia, and be intimidated by the prospect of saving Planet Earth, and that is a 'beaut' of a thought don't you reckon.

NEW 443. Didgeridoos and Kangaroos:
There is a lot more to New Australia, as it was with the old Australia, than didgeridoos and kangaroos. There are no

worthwhile reasons for clinging on to touristy gimmicks when the planet is in need of saving. The fact is that tourists will see what they want to see, and no amount of persuasion will take their eyes off the tacky. Tacky is more entertaining, yes, but it won't help save the bloody planet will it.

Drive north out of Melbourne, and seek out the bush country of old, the dry brittle areas of land that hitherto had not supported much green and animal life, and be prepared for a surprise.

After driving for about twenty-minutes it will slowly dawn on you that you are in a different Australia, greener, and with more trees. You, my friend, are in New Australia.

NEW 444. Land and lake, sand and water:

Being new is new, even to a politician on the scrounge this is a bland kind of thing to say, but it's the bloody truth. Just saying some t'ing is new does make it new, and no amount of bull manure will change that.

However, there is a redeeming feature for the politically minded. If the statement of newness is made on paper, and signed by Fred Jerk and Tom Shitface, that newness will take a long while to dissipate, and many discussions will take place on how to make the conflagration persist until the next round of electioneering is to take place. At that juncture, Fred and Tom will piss off and leave it to Mary Jentle and Sophia Thawt to sort out.

In the mean time, the rivers will continue to stay dry, stay dry, stay dry, 'rust', collapse in a heap of dry mud and sand, or trickle with water when there is a 'w' in the month, and no one will give a shit.

It is really a lovely thought if you are just embarking on a political career: 'I have made it my business to look into the anomaly of dry rivers in our country, and I will not rest until the problem is solved'. Meanwhile, Fred and Tom come back, and all is noted on paper sheets that are normally for other uses, and the land remains the land, the lake is no more, it never was anyway, and the sand and water fail to mix. However, the new politician makes a face-saving statement and becomes famous for caring for the dry rivers of Australia.

NEW 445. Worthy of attention:
It takes time for a new venture to gain recognition and support, but with New Australia, the recognition had always been there, fully worthy of attention, but smothered by old outdated ideals. 'New' or 'Old', it is still Australia, the sweet land of people who are fit and able, and ready, to lead the computerised button pressing folk of this magnificent planet to a new recognition of what our 'global space vehicle called Earth' is capable of. Planet Earth has been supporting life for millions of years, it is an ideal home for us, the only home for us, and if we can become less dilatory at working with Mother Nature, it will

continue for millions of more years. It is worthy of our attention. Be Australian, and stop faffing about.

Chapter 31. Homeland and People:

NEW 446 The importance of the Future:

One minute from now is the beginning of the 'Future', even if that one minute looks like being shitty it is still the beginning. It is you who can change that minute and build on it.

If you leave it to others to shape the new future, you will get their version and not something you visualise.

NEW 447. Confidences in each other:

Having a friend or companion is life drenched in confidences, confidence in each other and confidence in the life you lead. Widen that reliance to others around you over a larger area, and the confidence becomes more important.

Moving on to an infinite national level and the whole country becomes stronger, and life in Australia is easier by a huge amount for everyone.

NEW 448. A salad cream future:

A new land of greenery, grass, flowers, food crops and trees flourishes where dead sand and stones once lay in a earthly death, and it doesn't matter that the rangers play a didgeridoo, and paint their faces with salad cream?

Folk took what is on offer, and allowed the good folk of the future in Australia to reap a wonderful harvest.

NEW 449. Boots and bush hats:

Strength is not measured in physical abilities alone; there is strength in the ability to accept change. It ii this strength that is making the new Australia work for all, and it is this strength that has made New Australia and New Australians the envy of the rest of the world.

Boots are polished and hats are brimmed to perfection, to match the New Australia and its efficiency of living.

NEW 450. The gift:

She had struggled to care for her family, hardship had not been a stranger to her, but in New Australia, the gift she had always had in her heart came to fruition.

No one knew she had the gift, only she was aware of it, but she did not broadcast it, not in a deliberate way. However, her presence of mind at many difficult junctures in her life came to be proof of the gift, the shared gift of life every mother gives to her children.

Chapter 32 Authenticity:

NEW 451. Atom bomb at Maralinga:
The dictionary definition of an atom bomb:
A type of bomb in which the energy is provided by nuclear fission. Uranium 235 and plutonium 239 are the isotopes most commonly used in atomic bombs.
It is probably worth noting that the Maralinga bomb has left a sick legacy that will creep on for a very long time.

Author's Extra Note: Atom Bomb: A weapon that disregards any collective deployment, its only divisive 'aim' is to destroy indiscriminately – babies, children, mothers, fathers, grandmothers, grandfathers, and every animal young or old. It then continues to inflict havoc among the living for probably hundreds if not thousands of years.
If folk were to make it their business to destroy every nuclear device on this planet, they would be known as do-gooders by the country with a nuclear weapon, and peace envoys by the non-nuclear country.
Sadly, there are only two ways we can determine which definition and explanation is correct, nuclear war or non nuclear peace.
Will New Australia be the first non-nuclear country to follow the peaceful way?

NEW 452. In the economic sphere:

A new generous welfare state is now one the New Australians are justly proud of. No one is left by the wayside as the elders cogitate together and make life easy for all Australians.

The after effect of the end of the global warming danger, the country's finances and new business and social ethics gathered pace, with true liberation and tolerance.

Old businesses and new are flourishing in a true accord of favourable conditions, and the New Australian economy is stronger and more competitive than ever before.

NEW 453. The demise of the 'social scene'.

Stand or sit in front of an old work of art and relish in its antiquity. The soft tones of the painting will give a hint of the artist's view, of his or her take on the lives of the folk in the painting, for you to see and enjoy.

With each passing generation, the viewers and what they see changes and a new social scene replaces the old. After once advocating as resembling life as it was, the painting seems to move away, and the demise of the social scene is allowed only to hint at what was once a near true impression.

New Australia does not signify the old Australia, the comparison is non-existent, the old social scene, or order, of the old Australia is no more, and the future beckons with a warm smile.

Chapter 33. Rural Aspects:

NEW 454. Farms and farmland:
For the first 220 years from its rebirth on the 26th of January 1788, Australia had been looked upon as a new chance for farmers and their business of providing food for the nation and for export.

However, over-farming practises stripped the land of its chance to flourish.

Then New Australia was founded in another rebirth at the beginning of the 21st century.

NEW 455. Agricultural produce:
Whether it be a carrot or a cauliflower, or a sheep or a chicken, they are produced by human intervention, in a system of production loosely termed as agriculture.

Old Dictionary definition of agriculture, 'the science or occupation of cultivating land, and rearing crops and livestock: farming'.

Farming is an age old way of making provision for the continuation of human life on this planet, but where the old farmers managed bravely, sometimes in impossible conditions, the New Australian Agriculturist is using techniques that make the conversion of dry worthless areas of desert to green productive farmland. The hit and miss farming days are long gone, and we now can rely on a plentiful supply of food far into the future of this planet.

New Dictionary definition of agriculture – the cultivating of land to give us a guaranteed supply of food.

Author's Extra Note: As the Planet dies, the politicians stand to attention and sing Waltzing Matilda while being accompanied by a band of posterior licking 'do-gooders'.

By The Way:

NEW 456. Incompetence:
Not possessing the necessary ability means someone somewhere is causing hardship to themselves and others, they are being incompetent, which is not a New Australian attribute. Being short of the required level of skill will create havoc somewhere along the line, and delicate manoeuvring of thoughts and actions will probably only make matters worse. If I fall short with my writing skills, there will be many ready to be a critic, and want to see me heavily chastised, my delicate response to such is hard cheddar, what you see is what I wanted you to see and read, anything less would not do, not in my book.

If I fall short with my definitions, there will be those who will point the fickle finger of constabulary lawful condemnation. 'Now then, now then, what's this then?' My delicate response to such lawful bullshit is to ignore it completely.

NEW 457. Contagious microbes:
Waiting around the ecological corner are a group of microbes, organisms that have been dormant for thousands if not millions of years.

These micro biotic 'killer-creatures' have been in waiting, waiting for the right conditions, and those conditions are being manufactured by idiotic human behaviour, by humankind who are hell-bent on being right, even it kills them.

It makes no mind that you are at the top of your social tree, neither does being the wealthiest on your area mean anything. In the gathering micro storm, you will still be vulnerable.

How do you prepare yourself for the cataclysmic event?

(1) Stick your wet finger in the air and see which way the wind is blowing.

(2) Leave the country. Where to?

(3) Leave the planet. How and where to?

(4) Follow the New Australian way, and put up a solid natural defence by using prevention.

Author's Extra Note: Just one comment: 'Covid 19'. You can politically dance around that as much as you like, but it will not change the fact that thousands died because of a feeble response at the onset.

NEW 458. If it is not working:

It could be said that in a material society, if a thing is not working then get another one mend or replace.

Anyone in business to repair things would rather we have our dodgy thing mended, whereas, it would be expected that the person with a supply business would point out the benefits of buying a new replacement.

However, if the 'thing' is a new country, then if it is working why mend it. On the other hand, if the new country is as the 'new; signifies, it is beginning afresh, then new or more modern things might be the right way to go.

Please bear in mind, new usually means better, not every time I grant you, but mainly so.

All the previously mentioned is definitely not rocket science, it is more of the amalgamation of thought and practical application. Eh!

Author's Extra Note: Now I consider myself to be privileged, and I am thankful for the opportunity to be Australian, even if only by proxy.

NEW 459. The New Australian flag:

The last vestiges of the old ties with Britain and Europe flutter away, as the New Australia flag wafts purposely but gently in the breeze, at the very apex of the town hall.

Far below in the street, the people go about their day just has they have done for many years, either individually or in a collective descendant fashion.

no one looks up at the flag, and no one pays much attention to the town's new emblematic symbols or insignia.

However, flag, symbol, or insignia, New Australia is lately different and leads the way forward, the people united in a cause to cloak the world in decency. It is hoped that they will succeed.

Author's Extra Note: If what you have read falls short on your expectation, I humbly apologise and wish for you to do better.

Chapter 34. Any Other Business:

NEW 460. Keeping 'Stum':
If you say nowt, do not make noises above normal when it is too late, that's about as much use as a plastic frying pan on a campfire in the middle of nowhere, it will make a stink, but it won't do owt for anyone who happens to be hungry.

NEW 461. Time and Motion:
As the hours move on, the direction of the folk in Melbourne changes, the numbers gradually dwindle, and a kind of peace mixes in with the coming of evening and its slow drift to the dark of night.

NEW 462. Understanding through instructive news:
When report writers of the first European settlers of Australia made their assumptions, they made many errors of judgement. As a result of this, untrue records of events, and of the 'Blacks' as they called the Aborigines, people of the future were mislead, ill informed, and terrible events took place.
Since those rough times, we have learned to sift through the mishmash of those early reports, and enabled a better understanding of Australia's recent history.

The New Australia has been forewarned and forearmed, to make misinformation, or opinions disguised as news non-effective, and life has become easier for all.

The information now being passed on will help create a fuller and a more sensible understanding, and New Australia will be stronger for it.

NEW 463. Rotational influences:

'What goes around comes around'. How many times have I heard that, and how often has it been true? I have no bloody idea, and I have not the slightest interest in finding out. What is more important is that we understand that we live on a planet that spins at a fast rate.

In the past, humans have farted about making matters worse at every chance, while completely ignoring the obvious excellent opportunities to make life on Earth idyllic.

It is a farcical influence all of us are aware of, but within living memory, there has been wasteful and disgusting wars, and attrition, to make one race of people higher or more relevant than others, with smaller conflicts to make borders more exact seemingly worth the indiscriminate loss of life. All these irrelevant actions have made matters worse in every respect, but the humankind will continue to be abrasive to one another, to the point of self-destruction by atomic armaments, weapons that every man, woman and child knows will make this planet a truly dead one for a very long time. That is why New Australia has to be successful.

Future Australia

by

Leslie H. Harvey of Kimberley

Future Australian Social Humanity:

FUTURE 464. The first step:

Ucharonidge NT

Take a long slow deep breath... The air you are breathing has go to last this planet a long time, a very long time, a time span of time spans of time spans of a long time, so we should look after it, yes?

...

My dictionary tells me that fantasy is 'imagination unrestricted by reality'. But I say boo to that, and to any half-hearted attention to overcooked grammatical correctness, and change fantasy to fact, 'an inescapable truth'.

...

I suppose there could be a sense of trepidation in the notion of a utopian society; of the gaining of it, but it is perfectly possible for humans to achieve such a perfect society, the Aborigines did, some thousands of years ago, but to them it was just the normal way of living.

Then came the folk of piety from the Northern Hemisphere and they in their 'godly goodness' really made a mess of things for the Aborigines.

Ah, but, that is all in the past, and maybe that is best left in the past. What matters now, is the future, Future Australia.

Utopian society: A perfect or ideal social existence.

FUTURE 465. The Rainbow:
Derby TAS

When wholesome comfort is needed, and wished for, the reverse of all things that make life uncomfortable is what is required.

To give an example:

We all know what 'debt' is, an obligation to pay or perform. It is this dogmatic old-fashioned 'dealings' that used to make life rough, sometimes so rough that individuals took the clean way out.

Thankfully, debt of any kind does not exist in Future Australia, and not one person is in the purgatory of owing. Clean social equity is conducted in a fair and equitable way, and even though some folk are wealthy, wealthier than others are, their wealth is equalised by the Future Australian ethics of equality. The sensible cohesion in society, the interaction between one person, or one clan, or another, make it impossible for there to be any room for the need to forgive, and there is definitely no room for bigotry.

FUTURE 466. The useful and the good:
Havilah QLD

I do not know of anything or anyone being bad and useful, maybe I have been lucky, but I am sure there are the useful and the good, and that these far outnumber the bad.

To say that Future Australia is complete with a full compliment of just the good and useful Future Australians would probably be stretching credibility a wee bit too far, but I am sure certain that where there are bad, these are weeded out and brought to justice blooming quickly, and somewhat strictly.

Suffice it to say, I am confident that 99.9% of Future Australians are of the good and useful kind, and that the people of the world need to be the same, and be concerned about the expectations, and the seeking of the truth about this planet of ours. Not the political kind of 'let's be clear about this', but the good honest downright kind of truth we should all be proud of, at last.

No bull, if you are a Future Australian, you are of the best, you have to be, and all non-Future Australians should be pleased to live on the same planet as you, the same safe and peaceful planet we all now care for, at last.

Author's Note: Aborigine art and culture are not of a specialty with an Old Australian twist, they are Future Australian, good and useful.

FUTURE 467. Habitations of glory:
Milparinka NSW

Just a dip in the land makes it so simply a basic place to live, but to the people who live in the 'dip', they who inhabit the place, it is their normal place of residence, one of very many in Future Australia, and it is comfortable.

The Aborigine people of many different tribes had always lived in practical comfort, in sometimes, to the Europeans, very difficult circumstances until the end of the 'Old Australia' and the beginning of the 'New Australia'.

In Future Australia, the style of their places of abode has been upgraded, and made suitable for habitation by all Australians.

In addition to the dwelling upgrade, with the villages and towns of these fully integrated Future Aborigine style dwellings being situated in valleys, each community is complete in a unity somewhat dictated by location, and the Elders of these communities have formulated a wonderful fair and square multicultural social unity.

FUTURE 468. We never more will falter:
Newcastle NSW

The title of this section is akin to the first lines of an anthem, and I will try to live up to that, if only in writing style.

As life in Future Australia progresses to even better things and the people keep faith with the new ideas, it is easy for me to sit here and write of a Future Australia yet to come.

But when you are there in the flesh, and in the sun, then to be together has to be a united endeavour, and the maxim or motto is, 'we never more will falter'.

Then again, thinking on a wee bit, if minds are made up, Future Australia cannot fail to be the future, and a blooming good one, so don't mess about moralising, and picking out the grammatical mistakes, get on with it, get easy and true with the planet that is your home, our home.

Author's Extra Note: A new dawn my hacienda, it's a fresh start with bells on, so get ringing.

FUTURE 469. Time which we take on trust: Nullagine WA

Being completely mechanically logical, being late means you ain't going to be early, but if we think of the Time being Tuesday, and it is still Tuesday when we get to where ever it is we are going, you are not late or early, so to some extent Time in this context is of no significance.

Allow me to put it another way: if we declare the Time to be 'this week', we should have to be some kind of ninny to be late, would we not.

However, who actually does give a bent diddly? Time based on a mechanical division is irrelevant when you are living a simple natural life, based on Mother Nature's Time, which is planned in a scale of millenniums of seasons and the passing of night and day.

The pinpointing of Time by mechanical or electronic means only serves to add stress to a natural existence of motion, in a tiny part of an infinite universe.

Time, which we take on trust, is beyond our control. We can calculate Time, but we cannot control it, but we can always trust it to be on time.

Chapter 35. Nice and easy does it.

FUTURE 470. Is and must be so:
Mingilang NT

I had been finding it difficult to recognise the distinction of who is Australian, or to be more accurate, Future Australian, and I have lately come to admit to myself that the one and only distinction that sticks with me, is that there is no distinction. Either you can think about that for a while, or you can just let my supposition carry without objection.

If anyone can come from anywhere in the world and become an Australian, then it must be that an Aborigine, who are Australia, is and must be a Future Australian, or my name is 'wooflungdung' and I come from the Planet Mars, but only at weekends.

Okay, yes, I know, it is perhaps easy for me to sit on my super-cool mobility scooter, watching all world-kindred folk, of all ages, walking towards me and nodding, smiling, or saying a good day to me, or coming from behind me and having the decency to turn and wish me a good day in one way or the other.

This courteous attribute might not be just a Future Australian one, but it is a good show of how social interaction can cross all kinds of malingering taboo barriers.

My one sincere wish is that the last remnants of the barrier between Aborigine folk and other Future Australians will be smashed down and made to disappear for good.

No preaching, but heck, isn't it about time we grew up?

FUTURE 471. Cancer:
In a quiet place out in the bush:

I just don't want to write a 'poor me' anthology, but I can, and I suppose *I must* associate this with the experience of others.

In my many years on this Planet, cancer and its devastation has been all around me. Family members and friends, and acquaintances, all good folk, young as well as old, and all devoured by cancer, devoured without any evidence of mercy.

...

There has never been a magic wand or a magic formula, but there has been, and for many thousands of years, an Aborigine preventative remedy, one that had been ignored by the 'diplomatic quasi religious' medical profession, and the profit motivated pharmacy conglomerates. Note: 'diplomatic' is a polite way of saying something that is deserving of a stronger worded denunciation.

...

The computer generation's digital expediency of pointing us all in the wrong direction must have been one of humankinds' biggest confidence tricks, error upon error, and the whole world has been bamboozled into accepting it.

I will be back to this later, and if you are so minded you can jump straight forward, but I would ask you to remember, life and cancer are partners, when you are dead you don't get cancer do you.

FUTURE 472. The Future Australian Association of Ancestors:

Malbooma SA

Future Australia is not the only place on earth that has had a short ancestral history and an ancient dynasty. However, there is something unique in the Future Australian association of ancestors, and that is, as far as I can diagnose, the quiet supposition of one section of the society over another, and then the gradual inclusion of that same section into society.

It is the culmination of change creeping into the different sections of the population in a clear drip-feed of common sense, a strengthening that becomes apparent at every annual gathering, with each section missing the first bus, to then just catch the last bus, and getting to the destination in the end, without anyone having to explain their delay.

Author's Extra Note: It is not ideal that we should understand the reality of a situation, but it does help.

FUTURE 473. The dance of the seven knotted handkerchiefs:
Balanthria NSW

Author's Extra Note: I may be a wee bit biased, but this section is my favourite piece of prose, if it can be called prose that is.

It was a real Future Australian day, warm to hot, a clear blue sky, very little wind, with seven old codgers sitting on a couple of benches out front of the Goat's Head Hotel, each with a cool drink to the hand, and a knotted handkerchief on each head. The 'seven' were a group of Morris Dancers from England, and, of course, they were complete with bells, buttons, belts, braces and a scarf, and a 'wavy' white cloth each. However, their musician, an accordion player, had gone on what might be termed as a walkabout somewhere, he was nowhere to be seen. Then one of the hotel's regulars came out and asked if the Morrismen were going to give it a go.
After saying that their musician had gone off somewhere, and telling the regular that without music they could not perform. The regular shouted through the open door of the hotel. "Eh, Brian, we want some music out here, mate."

A couple of minutes, and out walked Brian, an Aborigine fella dressed as a cowboy fresh from Texas, and carrying a didgeridoo, to be then joined by two other didgeridoo players in similar 'country and western' garb.

At first, the droning of the didgeridoos did not seem to have the right connotation to Morris dancing, but when a woman came out, banging on a tin lid with rhythmic dexterity, the Morrismen began their routine with artistic vigour, and all was recorded for prosperity. The first known display of Morris dancing ever performed in Balanthria, or maybe in New South Wales, or even in Future Australia.

FUTURE 474. The tree is not the fruit Part 1: Mortlake New Town VIC on the 25th of April 2071.

They died for Future Australia and for Future New Zealand, and for the peace of the Future World.

They were the good grown youth-of-fruit from the family trees of Australia and New Zealand.

They left a legacy of continued life for all of us.

They gave their lives for that legacy.

(Not an extra day at work, or a volunteer day in a charity shop: no, they died, donated their lives for us)

They should be always remembered, every day, every one of them.

They; they are the ones to be remembered.

They of the family trees of Australia and New Zealand are in the future.

They are the fruit that made the tree.

We should not let them down.

We will not let them down.

We in our turn will be remembered.

Leslie H. Harvey 25th of April 2021.

Chapter 35. Motivated Driving:

**FUTURE 475. Driving in Future Australia: (1)
Karradale WA**

Author's Extra Note: If to gripe is to exercise the right to save lives, and be rid of the ridiculous Motorhead drivel, then gripe we should, until the Motorhead is dealt with once and for all.

What it the definition of a 'Motorhead'? A pillock, and a blooming dangerous one, a semi-autocratic road user who considers the road ahead to be their personal property, and the rules of the road to be for others.

Whereas, a 'Knight or Dame of the Road' is someone who drives according to the law, the commonsense rules of the road, in a safe and courteous manner at all times, and shares the road with an economy of dexterity akin to artistic generating of fine art.

Of the two distinctive types of driver, it is the Motorhead who will locate and hit the wall. However, before the pillock of a Motorhead meets the inevitable wall, the Future Australian Police have a legal preventative measure instigated by the Elders, and one that works every time; they take the car away from any miscreant at every opportunity and scrap the blooming thing.

Any cries of civil liberties or any such gibberish are just whimpering blooming hogwash, and written or verbal condemnation of such remedial action is of no significance and is be ignored.

Furthermore, there is no way any Motorhead 'worthy' of the name will risk having their beloved motor vehicle taken away for scrap because they 'failed to signal in good time', or they 'did not give way at a give way sign', or they exceeded the speed limit by a couple of kilometres per hour.

Future Australian Elder traffic laws save lives, and *every* motor vehicle is driven, parked, and perfectly cared for within the law in every respect.

Author's Extra Note: If you doubt the veracity of the previously mentioned, check the local statistics for death on the road.

FUTURE 476 Driving in Future Australia (2): Albony WA

Author's Extra Note: The only real epitaph was a jagged groove in the tarmac, and this was soon obliterated in a denial of reality.

Contrary to a popular misconception, driving a car fast is not a sign that indicates that the driver possesses great skill and acumen, no; the opposite is nearly always the case.

...

Take a pause here for a moment and think on:

These are all real events.

A child of 5 killed by a hit and run driver.

A mother and her two children, one 3 and one 4, all killed, trapped in a burning car after being

hit by a speeding motorist.

A young man of 18 lost control of his speeding car, hit a concrete post, and died at the scene.

...

They drive up and down the public highways and byways, the idiot pillocks sitting behind the wheel of a motor vehicle that has no conscience, it doesn't give a Mickey Fin who it kills, while the driver attempts to get where they want to be, and never mind the cost.

However, some of them would be better to stay where they are, to live longer and happier lives, in near tranquil stillness, and cherish the calm of knowing they are right by living the Future Australian way.

A car that is scrapped by the police brings joy to no one, but it does save lives, even yours.

Author's Extra Note: In a completely personal way, I say the best place for a powerful private car is in the garage. Break the 'neck' of a car and it can be scrapped. It is a different issue if the neck belongs to the driver or some other person, young or old.

FUTURE 477. 'BDMD'. A short history lesson: Pannawonica WA

Author's Extra Note: I am a wee bit paranoid, and that I cannot deny. I detest those who say they can drive when it is obvious they cannot

You can cringe, whisper a blasphemous remark, or you can turn the page, or even throw my words away, but I am going to do it anyway.

Bad driving of a motor vehicle on the roads and tracks of Future Australia is no longer tolerated, and of that I am mighty glad. It is not just bad driving, it quiet often means DEATH. Bad Driving Means Death 'BDMD', was a real menace that was ended by Elder decision of the scrapping of vehicles used so as to offend decency of life.

(1) Yes, I have gone a wee bit intimating, (not the actually pointing the finger) but if by some miracle I can help save a life, then cheers to the end of BDMD.

(2) Yes, as I have intimated, I am a wee bit paranoid about bad driving and its cronyisms affecting the young and old alike, male or female, not to mention the harm being done to the sweet Planet we live on.

(3) No, bad drivers do not have a leg to stand on, although, when caught, they will be standing on two legs that may be a wee bit wobbly.

(4) Now then, take a slow look around and listen. Those dust particles in the air, and the noise constantly in the background, are the after-effects of bad driving in track and road traffic. In Future Australia this dust and noise is hardly a bother, confiscation of a vehicle does have a steadying affect on driving techniques.

(5) Yes, I know, quiet tracks and roads are now taken for granted, but be aware, there is always the lingering idiot of a BDMD driver somewhere near you.

Gripe over.

Chapter 36. Global:

**FUTURE 478. The heart of a Global Writer:
Mildura VIC**

A writer can put his or her heartfelt thoughts on the paper, but 9 times out of 10, no one gives a flea's left eyebrow. That is a salient fact, and one I have experienced many times.

However, I will not be deterred from writing what I deem fit, and if my efforts in this way can help save the Planet, blooming heck, what better motive is there! If I fail, I will try, and try again.

...

The utterance of 'language', some folks call it swearing, is not so fierce when seen in the written form, and I make no excuse for such 'language' in my work, in fact, I am proud that my publisher saw fit to commit it to print as it is. Simply by its use, it makes 'language' and grammar of less importance, less important than getting the message across, that of saving the Planet.

This is a kind of indifference to scholastic or ecclesiastic condemnation, stems from my ability to see the wood in spite of the trees; the wood being the downright disregard for our Planet's survival, and the trees are the pillocks who try to instigate remedial work, 'work' that amounts to nothing more than a sparrow fart in a sandstorm.

There are many thousands of examples of political interfering correctness all over this world, and their non-effectiveness is the one feature that stands out on examination.

As I have mentioned many times, being politically correct is of no use whatever, and in my heart, I know I am doing the right thing for Future Australia and the Future Planet we call home. Future Australia is our one redeeming feature on a world that had gone amenity crazy, just get on board and let's be good for our children and their children, descendants of ours who would wish to live safely in the Future World, on our beautiful blue Future Planet.

FUTURE 479. Jack Kennet's favourite spot: Beetaloo NT

In all of Future Australia, Jack Kennet has a favourite spot, and it is right by the front door of his cottage/shack, just two steps out and three to the right and he was there.

The patch measured roughly 1.1 metre by 2.4 metres, and it is perfectly level. Weirdly, it is surrounded by a mixture of stones and low growing bushes, but not too close, and long grass, and weeds.

Between the sides, for the full width and length, there is the finest short green as green grass, with not one weed to blemish the exquisite oblong patch.

Aside from the foresaid, there is something else special about 'Jack's Patch'. Believe it or believe it not, it is the handiwork of the little folk, little folk Jack came across on many occasions as they worked on his Patch.

At the beginning, Jack would tell of his little neighbours, but as time went by in chunky years, and no one else had seen the wee folk, he ceased to talk about them to anyone...

The years past, and sadly Jack eventually died, and the following years of neglect saw his old cottage/shack crumble away to practically nothing but a few disintegrated wooden bits here and there.

On the other hand, I am happy to tell you that Jack's Patch is still in pristine condition, or it was when I visited the old place a few months ago before I sat down to write this. Other 'wanderers' I am sure have wondered at the small oblong patch of perfection in the middle of the bus country, just by the Mitchell Highway, wondered and then forgot about, but I won't forget, and I can't help thinking how good it would be if we all had our own 'Jack's Patch' by our own backdoor.

FUTURE 480. A huge progress:
Hahndorf SA

My writing is not about making friends, and, hopefully, I am not accidentally influencing people in the wrong way. But I am certain that I am doing my best to get the people of this world to wake up to reality, to forego the materialistic life they have indulged in, and have become addicted to without any caring thoughts for the Planet Earth, our collective home.

Author's Extra Note: I use the 'Planet which is our home', or a similar phrase, many times, and I am not about to apologize for this. The Planet of ours is worthy. If it makes you cringe that I should be so naive to think I can make a difference, then, my dear reader, cringe away, while our Planet struggles to survive against our stupid human ignorance.

In all honesty, it is possible that I don't have a gnat's whisker of a chance, but I know I have to try, even if I am the only one in thousands to do so.

However, there is a mind-settling dilemma in all this. I have considered that with me possibly being the one in a thousand or a million who thinks it worthwhile to try, I could at least relay one certain fact, that in every thousand there could be one like me, making one thousandth of a difference. However, Future Australians have proved that just 01% is enough to save a country, and consequently a world, a Planet that deserves better treatment.

If you are one of the 'nought-one-percenters', good on yer, and do not give up your place. You are okay, wherever you are, and if you do meet another of our one-percenter colleagues, that in itself is huge progress.

FUTURE 481. A lucky escape:
Esperance WA

Author's Extra Note: Downtrodden, 'issed off to the back teeth with neo-political bull manure.

That was the lot of the Greenpeace folk of the 2020s, good individuals who were aware of the real state of our home Planet and its path to an Armageddon of fire and flood.

Thanks to the Australians, Greenpeace members' worst forecast has been proved to be wrong. It was the Future Australians who called a halt to the human folly of damaging the environment of a beautiful Planet Earth, and it is the Future Australians who now lead the way, with Greenpeace acknowledging that we have all had a lucky escape.

FUTURE 482. The benefits of a shower (or unadulterated bullicose):
Auburn SA

Author's Extra Note: When the full force of logic hits you, do not try and dodge the issue, just 'grasp it with both hands, and make Logic equal Commonsense.'

If you bathe in warm water on a hot day, it is likely that you will be warm and wet, or even hot, and then sweaty and wet.

However, if you take a hot bath on a cold day, it is more than likely that you will think it is blooming freezing when you step out of the water.

Now, with reference to the previously mentioned rhetoric that is within a gnat's whatsit of being pure bullicose, I have discovered that by not having a bath or bathing, and keeping quite still, especially between the evening of one day and the early morning of the next day, you can cool down naturally on a hot night, and stay cool.

Therefore, if you are going to take a bath, or bathe, it might be best to make sure you do so on a day when it is neither hot or cold, and even when it is what you might call alright, and maybe you should think more than twice.

Of course, you can ignore all the above, and be a proper Future Australian, and think on... act on...

...and gain the benefit of having a shower.

FUTURE 483. A blessed plot on this Planet: Etheldale QLD

Here we go again, put the imaginary poetical flag out and let it waft and flutter in the breeze of appreciation, if there is one, and as you think of it, spare a thought for all those creatures that are reliant on us, they cannot speak. The wild animal population who have no say, and probably couldn't give a gnat's whisker for any genre or flag, imaginary or not. They don't even know that they are relying on us.

FUTURE 484. Yours is the Planet and everything on it: Quamby QLD

Obviously, 'Yours is the planet and everything on it', is not meaning you personally, but in a philosophical sense that a Sci-Fi enthusiast would probably understand. Nevertheless, you *are* entitled to everything on the Planet, and the Planet itself, but it is a privilege you must earn and deserve by your efforts, not just by your 'rights', whatever you think they are, or by the fact that you are rich beyond dreams. Money is just that, money, it is not a key to the Planet; Mother Nature and Mother Earth have those keys, and they ain't about to share them with an old-money-bags-senator or political dollar counting bootlicker.

...

The sky above Future Australia is the same sky that was above the Old Australia of the 18th century, but with at least one massive difference.

The sky of the 18th century was more in tune with the Planet, and what was desired of it. But the sky of the 19th and 20th centuries had to contend with pollution on a 'grand' scale. But now we are in a much better situation, with a return to the 18th century conditions. The earth of Future Australia is caressed by rain; clean rain, and warmed gently by the sun, both in regular adequate amounts. This makes soils and plants of good health, animals with adequate food and water, humans behaving as they should without any form of stress, and with Mother Nature and Mother Earth in a harmony of influential ecstasy, and creating and maintaining Planet life in a fair and happy state, in fact, utopia, at last.

Author's Extra Note: There'll be peace in the valleys some day, some day soon.

FUTURE 485. Age is not a timetable: Horrocks WA

There may be fresh cool water trickling in the stream by your feet, and the gully the water is flowing through is probably as old as the hills, and, most likely, you are not the first to stand where you are.

...

Age is not a timetable; it is just an indication of how things are.

...

In Future Australia an all year round constant stream of cool clear water is now accepted as a normal land feature, and the fact that it contains a good flow of water every night and day all year round, every year, is also accepted as being of the normal state of affairs.

Furthermore, dear reader, Australia of the Future is also of the 'Fortunate', and the people of Australia, all of 'em, are Future and Fortunate Australians, and, by 'adoption', the whole world is of the same ageless fortunate viability, thanks to Future Australia and its good people.

Author's Note: I do not give a monkey's left dodah what folk might think of me, or of my writing ability for that matter. What is important, is that the world follows the lead of the Future Australians, and save this beautiful Planet of ours for the good folk of the future, bless 'em all, and why shouldn't everybody be fortunate, and I mean everybody.

FUTURE 486. In good hands:
Hallett SA

If there is safety in numbers, how about the millions of Future Australians; and each one is as safe as they come. It could not be better, and we, and the Future Planet, are safe in good hands. Of course, there are those of the world who may think my prophesying to be just prophetic codswallop, and unworthy of any academic consideration, that, my dear reader, is as it is, they, and maybe your good self, can think what they or you blooming well like. However, let me gently point out the following:

1.At the time of writing, May 2019, the beautiful Planet Home of ours was under threat of extinction by human destructive actions.

2.The encroaching extinction, was by the human population in a completely selfish denial of the obvious.

3. The Australian continent is isolated from the numbskull administrations of the rest of the world, and with the adoption of the Aborigine Elder system, Future Australians can help the world to safety.

4. The Aborigine peoples have been Australian for over 80,000 years, 80 millennium, and we modern earth inhabitants have been on the Planet here for just 2 millennium, and in Australia for less than 300 years: we know nothing at all it would seem, and until the onset of Future Australia we were too blooming pigheaded to recognise the fact.

5. The original Australians, the Aborigines, got it right, and we got it drastically wrong, and we really churned it up for them.

Of course, you can take a pill or a powder, or sniff a weed, and think of your personal comfort, and allow our descendants witness the Plant destroying itself in a fully defensive strategy.

Author's Extra Note: You don't have to like everything, unless you're running out of time.

FUTURE 487. What the word did make it:
Ongerup WA

Note down a word, any word, on a page in a notebook, and put 'Future' before it, and then put the notebook to one side; if you are using an electronic note taker, save the notation you have just made, and forget it.

...

Every sentence ever made begins with a word, and that is the nature of things in a literal sense, but starting with the word 'Future' will create a new prospective to your writing or vocabulary, and in such a way to possibly change your life - don't believe me?

...

Go back to that notation you made and look again... *You* wrote that, and it is you who can sense the change.

When I first wrote 'Australia', some long time ago, my mind's eye was resplendent with pictures of a street in Melbourne, a dry creek bed near Bingara, an Aborigine sitting quietly, and kangaroos grazing. Then, after placing the word 'Future' before 'Australia', all the pictures changed. The Melbourne street was quiet, free of traffic, the creek was full of water, the Aborigine was smiling calmly in a clean and ordered village setting, and the kangaroos were grazing on soft green moist grass. I really like my view of Future Australia.

FUTURE 488. The acorn is not the tree:
Jerangle NSW

We did not construct the planet, 'tis t'other way around,' and if Future Australia is to be the utopia we all should have, and share, then we should stop our indelicate messing about.

An oak tree might from a little acorn grow, but that does not make the acorn the tree, Mother Nature takes care of that, and Future Australia is going to taker care of all us little acorns.

If you can't take my word for it, come with me to the 26th of January 2071, and relax, it is quite possible that you are going to enjoy it.

...

Australia Day in the year 2071 looks like being a fantastic day, with sunshine forecasted for the whole day all over the country, with a temperature in the region of 35c, with a breeze or two, depending on where in Future Australia you are.
I am in the small town of Jerangle in NSW, and I am sitting in a Breggs 'coffee and cob', a food-store with seats, at 0815 in the morning, and it is raining lightly, so much for the forecast of a sunny day, but it is warm.
The sliding doors of the food-store are open wide, with an overhanging roof protecting us from the precipitation.
My bacon baguette with brown sauce is going down a treat, helped by a large white coffee that is more black than white, and...
Wait on a minute... yes it is... it's the sun, it's beaming down its warm to hot rays, and the morning is improving in leaps and bounds, *it is* going to be a good day.

FUTURE 489. When we hear the fountains: Woorabinda NSW

It was raining, not in the get soaking wet quick sense, but in a few 'spots' here and there kind of way, and I did not mind one bit, in fact, I would like it to rain a lot more.

My proclivity for rain, and plenty of it, was not born of a desire to end a drought, those sorts of conditions were no longer relevant, they belonged to the Australia of the old days, whereas my unspoken wish for rain was because it brought great joy to Woorabinda and the countryside around.

When we here the fountain of rain supplied by Mother Nature for free, we are a long way from the syndrome of a dry river bed, and river banks of dry weedy, sparse feeble grasses, hardly a tree at all, and low 'scratchy' bush growth.

Now, in Future Australia Woorabinda, there is a full river of a constant flow of sweet freshwater, with trees in abundance along the banks, tall and fine-looking trees, with wide reaching branches of green leaves, and beneath the branches are the flowers and good green grasses in an easy full state of growth. I could almost feel the joy of the natural growth around me, and I could feel the rain, making it as though I can hear the fountain of Mother Nature as she feeds and caresses the land of Future Australia, I know I am in the right place, and a good right place it is.

By the way:

FUTURE 490. Bubbles that shine like beads:
Dargo VIC

The water in the stream was crystal clear, and at a point where a trickle had become a tiny waterfall, the cool clear pool formed beneath the fall of water was bubbling, with up-drifting bubbles of air that resembled shining beads.

These beads of air were a continuous response to the falling water from above, a constant supply from a landscape that had once been dry and bereft of any green life, but is now full of life, green and ambient in the gentle way of Mother Nature. To think of it as otherwise would seem to be a denial of the Future Australian association with hope and satisfaction of natural life.

Chapter 37. Farming:

FUTURE 491. Watch the green fields growing:
Dargo VIC

It is said that watching grass grow is a non-participation hobby. You can watch it and wait, but you will more than likely think 'bother to this' before any sign of grass being grown.

However, in spite of the obvious drawback to your new hobby, I can assure you; I have it on good authority, Future Australian grass does grow, and at a phenomenal rate.

With a combination of good soil nutrients, with sun and water in adequate amounts, Future Australian grass is a world-beater, an excellent pasture crop second to none.

Of course, no farmer or agriculturist worth his or her salt is going to sit and watch grass grow, that would be tantamount to dereliction of duty by default. Do not watch it, use it, get to know it, and grow it, nurture it, and enjoy it

Now that Future Australians have the drift of how to harness the sun's power, and the life giving liquid inductions of water, the world can celebrate a lucky escape, and when the next rainbow appears, you can rejoice in the thankfulness you have for a forgiving and bountiful Mother Nature, a Future Australian Mother Nature.

FUTURE 492. Come wind, come 'weather', come sun, and come shine:

Jingermarra WA

There is an expanse of wheat fields and farms, with some of the farms predominantly involved with sheep. There are other livestock farmers, and these are concerned only with kangaroos. However, whatever the farming practices, all the animals are treated with respect, even though some of them are destined for the meat market.

In order for me to be accurate in my note taking, I made the conscious decision to recognise the human tendency to eat meat in a 'natural' form of continued existence, for the human race that is.

To be perfectly honest I do not know very little about livestock farming for the meat market, but it is a well-known fact, that in Future Australia, kangaroos make for good farming practices, their meat, so I have been informed, is excellently nutritious, but so is a banana. But then...

Nonetheless, come wind, come weather, come sun, and come shine, the kangaroos are on the farm, and still survive in the wild, with the country and its people better off for it.

There is controlled grazing in the new land of plenty, and the land remains green and fully productive, and more so every day, with no risk whatever to returning to the dry barrenness of sandy plains that were once the blight of Australia.

FUTURE 493. Barbed and no thicker than a lace o' liquorice:
Smithville SA

All that separates thousands of square kilometres of Future Australia, is a few strands of wire, extra, extra long, barbed, and no thicker than a lace of liquorice, and maybe not so strong where it has rusted.

Whether it be historical or not, that wire barrier contains many tales of endeavour and misery, some of which are tales of long gone events of which no one will ever know about.

With the onset of the birth of Future Australia, the outback has shrunk somewhat, its edges less 'powdery' and more green, flourishing with natural life. Even though the barbed barriers are still in evidence, their validity is dwindling, and many folk regard them as being of no real significance, other than to define a boundary that no one really cares about anymore.

In Future Australia, a stone placed in a strategic spot is sufficient to make a boundary legitimate, nothing else is required, and life goes on peacefully as the land responds by expanding its green fruition further across the land, bringing the countryside to a new relevance.

FUTURE 494. Small sticks make big fires: Mangalore VIC.

Author's Extra Note: 'Freedom is always worthy of struggle, but the struggle needs to be worthy of the freedom'.

Believe it or believe it not, all of Future Australia is made safe by the use of fire. Whereas in the bad old days of the late 20th century, and early 21st century, when fire was a real uncontrollable and deadly enemy of Australia and Australians, Future Australians use a small stick to make a fire deliberately, quite often a big fire, but one that is under full control of those carrying small sticks.

Taken from the Aboriginal system of burning bush to order, the Future Australians have learned to create good farming land for crops and livestock, simply by starting a fire. In this 'new' way, all weedy and neglected land is cleared and converted to good use. This makes way for the farmers/stockpersons to bring about a monumental change to food production on Australian land.

A simple method of land management and one that helps the environment of Future Australia, to the extent that deserts of any great size are just a memory.

Forests have taken over from the spinifex barrenness, and surprisingly, the air is cleaner in the complete opposite of a fart-in-the-face conditions previously persevered with. We should not fart in the face of Mother Nature anyway, apart from it being rude, it stinks of a total disrespect of what is good for the planet.

FUTURE 495. Nothing in it for tears:
Kahmoo QLD

What's the good of crying over spilt milk? If you are in the process of making cheese, what's the loss of a litre or two of milk going to do, you might as well hold back the tears, and create a new metaphor choice such as, 'oh dear, I've spilt a few drops of milk', and continue with the cheese making.
In the great scheme of things, cheese making or not, a bit of spilt milk does not amount to much, and there is definitely no reason for tears.
Slowly but surely, as the cheese matures, the spilling of milk, or even milk itself, will not come to mind, the dairying intrigues have moved on; the customer is beckoning, and your cheese is destined for consumption in a happy dry-eyed Future Australia.

Author's Extra Note: 'In day-go-days, we did what mum and dad said, and life was simple. Now, we do what the computer says, and life gets complicated quicker.'
That is unless we go the Future Australian way.

By the way:

FUTURE 496. Hopes are not dupes:
Minderoo WA

...

Hope:
A feeling of desire for something, and confidence in its fulfilment.
Hoodwinking:
To deceive by trickery of some sort, and usually for gain.

...

In those day-go-days, Australia and the world were in a very precarious situation, the rising global temperature held a threat that people did not seem to know how to offset. Even if they did, they did not have the surety of get-up-and-go to get on with it, they 'ummed' and 'aahed' while the planet slipped towards a comma of death by fire, storm, earthquake and flood.
Then there came the clarion of a new hope in the form of New Australia, a new world leader in the eventual success over the so-called global warming.
However, before the new way of living could be fully established, the hoodwinking merchants could not let go; their investment in the 'stuff you' ways of living were too attractive to them.
They used dupes, especially political ones, to try and stall the common sense approach to life on this planet; happily, oh so

bloody happily, the dull heads got it, and consequentially, they failed in their hoodwinking of a whole planet, and now, even though we all rely on computer systems for guidance, we do so with easy common sense, New Australian common sense.

Author's Note: If you are not New Australian, and you find it difficult to accept their leadership, tough mate, without it, and them, you would be looking across a 'landscape' of muddy water, or fighting off the flames of an uncontrollable firestorm.

Chapter 38. Future Australian Living (Part 1):

FUTURE 497. The tree is not the fruit:
Mortlake New Town VIC
What happens when we actually pursue our dreams for our home? If rather than think that we can't, we think and believe that we can, then the dreams will come true in the passing of time.

...

If you change the notion of believing that 'the tree is not the fruit' to 'we are the fruit', your search for a home in Mortlake New Town will be soul satisfying.
Likewise, if all Future Australians think of Future Australia as the country that can and does make dreams come true, then their soul satisfying will be manifested in the outcome.

FUTURE 498. The smiling land of 2071:
New Lerida NSW
A Future Australian calendar of joy, not of days wasted in recrimination or finger pointing, but of days spent in good honest enjoyment of life in Future Australia.
In the village of New Lerida, for that's all it is, a village, the people, all 74 of them, live in a unison of inter-racial and non-

racist ambience that the folk of the district of the years ago would have thought impossible. However, this is Future Australia, and such an environment is the norm, and to be expected without a second thought.

FUTURE 499. Future Australia and a meal on the go: Wirrulla SA

I found a table for myself in the Robyn Cafe, an establishment that seemed to glow, with clean white linen on the tables, shining clean chrome on the equipment behind the counter/bar, and the unmistakable afterglow of the many happy customers who had made their acquaintance with the comfortable dining establishment.

It had been my intention to just take a quick snack and then depart, and continue on my journey of discovery, my discovery of the Future Australia I was beginning to accustom myself with.

However, my quick snack was forgotten when I perused the menu card. Why was it that such good food was available out here in what was once brittle hot bush 'make-do' country? The question made it s way through my wanderer's mind as the prim waiter approached. She was young in years, but, as I was soon to learn, her knowledge and attentive nature were of experience and not some little skill.

Within a matter of a few minutes, I was made to feel welcome in a kind and comfortable state of customer and waiter association. I was at ease as I was relieved of a few dollars for the meal of perfection that was put before me, a meal, I can tell you, that I enjoyed immensely, and in Future Australian cultural surroundings of excellence.

FUTURE 500. A friendly peace and a cup of tea: Burtundy NSW

'No matter how you struggle, or whatever is the bane of your life, you are alive'. Those words were said to me at a time when I was low, and although they probably did not make much difference to my situation then, years later, sitting on my own in a warm sandy spot under a gum tree, and away from all other divergences, the words finally rang true, or the thought of them did that is.

My life was now an older one, and I was in a free country, far removed from the stress of having to figure out whether to stay or leave. I am in Future Australia, a country that is a huge nature reserve, and a country where there are many friends and acquaintances and no strangers, why should I wish to leave such a place.

I have made myself 'a pot of tea' in a can, a billycan I suppose you could call it, however, it is just an enamel can with a lid, a can I fill with water and place on a small camping stove to bring the water to a boil. I then add some tealeaves and wait while it brews. To my mind, by this simple gesture of my independence, I fully justify being where I am on my own, while at the same time I am quite willing to make another pot of tea if someone would like to join me. I have no problems with that at all.

I raise my tin cup to you, and wish you well in this Future Australia and this Future World of good will and clean air living, and I am sure you will agree, there is nowt wrong with that at all. 'Sit thee sen dern, an' mek thee sen comfortable me owd mate'.

FUTURE 501. Adella Wilpena and the rain: Inglewood NSW

The old woman said that rain was only water, and she was waterproof. Common sense in a rainstorm, from an old woman holding an umbrella while she was sitting in a wicker chair, on a porch veranda, under a huge overhang of her bungalow roof - she certainly was waterproof.

...

Delicate by nature, and delicate with her words, Adella Wilpena had never minded the time to be different, but she was old, of that there was no doubt.

Her porch veranda was about as far as she strayed now, in her 79th year, in the year 2071, although she could remember the days when she roamed near and far in Australia, the Old

Australia, when there was the real threat of global warming, and the dangerous state of affairs she and Australia were heading for back then.

However, rain or no rain, it was different now, and Future Australia had made things oh so kind for her. No amount of comments by anyone would change her outlook, a clean and happy outlook that contained the flavouring of common sense, in a country overflowing with common sense.

Adella needed no explanation from anyone; she could feel it in her old bones that everything was just fine in Future Australia, just fine.

FUTURE 502. Season of mist and fruitfulness: Wambiana QLD

She walked slowly, in a contented way, her hands resting on each other across her front, and with a smile on her face that would soften the severest frown on another.

Her walk was taking her through a churchyard, with headstone markers indicating the 'presence' of those long departed.

The passage of time had left a delicate rustication on the environs close to her, and even though there were probably over two hundred headstones, the Future Australian attitude she had, helped her to see them in a strange but comforting way.

Even though they were lifeless beneath the headstones, they, even in their non-physical sense, were still part of Future Australia. As if to signify this, over in a corner, standing somewhat serenely, were a group of eucalyptus trees, their blooms providing a true glimmering indication of a good summer to come. It was to be a special spring, a season of mist and fruitfulness, and a triumph over death and its finality. With the fragrance of the eucalyptus blossoms engaging her, she moved her hands slightly, and as they resettled, the smile on her face widened. She was a Future Australian, and one who was experiencing an acceptable peace with all that was around her.

FUTURE 503. For evermore my dwelling place: Marlborough QLD

'It's where I live, and it's where I intend to stay'. So said Jimmy Jackson in 1819.

Of course, he was one of the original European settlers in Australia, and it was one of his descendants that told me the story of 'the old man' of the family.

The old man skilfully threw or chucked a boomerang he had been given by a native Australian, and he apparently said; "Where it lands is where I will build me a house", and he did, calling it 'Square One'.

Perhaps that is a wee bit too sentimental of me to write in such a way, but that's how it came across to me. Nevertheless, I can tell you that 'Square One' is still there where Old Jimmy built it, and from what a young Mr Jackson told me, I would reckon it will be there for a good while longer.

However, whereas when it was first built, the surroundings were just basic open bush country, and not a spot to be over fond of, the house now stands in the midst of a green and pleasant land of flourishing crops, and healthy livestock grazing, the essence of a domicile in Future Australia bequeathed from the past.

FUTURE 504. Hear and believe your own truths:
Old Coralie QLD

You can look in a mirror and see your own face in all its true definition, but will there be a reflection of the truth that is you? Even in Future Australia there are some untruths, misrepresentations that the individual will not except of themselves.

The sound or motion of your breathing is the audible and physical evidence of you, while the look in your eyes will tell a story that will match your breathing, and it is this story that will relate the truth of you to others.

Depending on your age and social standing, you do not mind recognising the truth of you if you are of a happy disposition. However, if the truth of you is reflecting a lie. That lie will be there in your eyes, if you can bring yourself to look closely, and

I mean blooming closely, not a quick glance and a mental nudge, but a long honest stare. If there is a lie, you will see it. If you can be true to yourself, any lie you might be percolating in your mind will be reflected in your eyes in the mirror in front of you, just look, take a right good look at the real you.

As the genuine Future Australia is resolved around you, your own thruth becomes a reflection of that national resolve, and this will serve to indicate your sweet life in Future Australia, an indication that will be, and is, reflected around the world.

Don't mess about now you are in the lead, you and your fellow Future Australians are taking the right action, use it, make it happen, and don't listen to the political pillocks, they are just a waste of Future Time, they do not belong anymore, and the sooner they recognise this the better.

FUTURE 505. Listening to the quiet of the moonlight: Indiana NT

The moon does not make any sound that the human ear can detect on our sweet Planet, and even though we can often see it in all its reflected glory, we cannot listen to it: or can we?

Our Planet of Earth has been so for a very long time, and others before us have looked at the moon and seen what they wanted to see

Whatever their mind and eyes told them, the old Australians believed in interpretations that were made to fit superstitions common at the time, without resorting to 'listening'.

Now then, if you are in the middle of a large area of Future Australian grassland, with not even the rustle of grass, or the sound of some night creature to break the cold and dewy silence, you can let your thoughts be gentle. You can relax, and allow for the moonlight's shadow casting to 'decorate' the area around you, no deviations from the past will influence you, and you can let it be a Future Australian tête-à-tête between you and the moonlight. Let your shadow and you be peaceful, and remain still as you stand in the quiet of the moonlight, on you own with no one to see or hear you as you listen to the quiet of the moonlight, and to the wonderful presence of peaceful existence in Future Australia.

You do not have to be some clever intellectual to recognise true, really true, human evolution as it happens, just be yourself, your real and true self.

FUTURE 506. Old are the woods:
Nindegully QLD
Youth and vitality do not last forever, and in my 79th year on this sweet planet of ours, I can vouch for such a viewpoint. Although my inner being, my real self if you like, is still young; around 24 years of age, I am what you might call young at heart. Whereas, my frame of muscle and bone is indicative of the appearance and feeling of old age... but, eh, sod that. I am sitting here in Gregg's Coffee Bar in Kimberley on the morning

of Saturday the 26th of May 2019, just three days after my 79th birthday celebration, and I feel special, Future Australian special.

With my hot mug of coffee on the table in front of me, I am reading and writing, and hearing other customers ordering their bacon roll, or just a cake or a biscuit, with probably a cool orange drink.

Through the open doorway, the morning sunlight of what could be a good day creeps in and the scene lights up before my eyes. To me it is as if I am in a tiny corner of Future Australia.

Author's Note: Attached to the front of my mobility scooter parked just outside in the sunlight by the open door, there is a small brown monkey doll, I call him 'Fred', and I am proud to introduce you to Fred. However, aside from that, maybe Fred and me will see you on Sydney Harbour Bridge one day, and you can buy me a coffee in a coffee bar in Sydney, Future Australia Sydney, my treat. I tell you that would really make my day. Until then, hears to you and yours.

FUTURE 507. Every settlement, hamlet, village, town and city:

Murgaenella NT

Where there is an old railway, there is was once a new one, with trains that had somewhere to go that meant something. Now in Future Australia, the old railways are new again. The people who live in the places, and the people arriving on the trains are happy and grateful for the trains. The travellers have a new desire to meet with new adventures by rail, and the residents are happy to meet the travellers who arrive on the trains. It is like a brother/sisterhood of travellers-by-rail who know what they are at is right for the rest of the world, and it is good and proper to be as they are.

There is a welcome for every visitor, and there is a place for each one in Future Australia, in every settlement, hamlet, village, town and city.

Feeling strange in a new place is the normal way of things, it is the ability to be Future Australian and to retrieve and hold onto your self-respect.

Author's Extra Note: Remember, all Future Australians have come from somewhere other than Australia, and you and they together are the Future, you are an essential part of Future Australia, and there is no need to be reserved about that.

FUTURE 508. Lowanna's proviso on life in Future Australia:
Coorada QLD

> *Music washes away the dust of everyday life.*
> *Berthold Auerbach 1812-1882*

Lowanna is lovely in her old age beauty, and in no way could anyone deny her the accolade, but she had one proviso on life, her grandchildren.

Her Australian upbringing, and that of her children and grandchildren, had made all the difference, with fairness more or less the family byword.

Now she was on in Future Australia, and she was experiencing a lifestyle she would have once thought to be beyond her reach. All amenities were geared to her ability, or lack of ability maybe, and there were no difficulties in living to a high standard, but her grandchildren really put the shine on things in a proper Future Australian way. They told granny how much they loved her, but could she please try and have a bit more get up and go.

Sadly, Lowanna had lost her get-up-and-go some years before, but she tried, and in time with some lovely music, she lifted one finger and one eyebrow at a time, in her version of a keep fit regime.

The grandchildren laughed, Judie laughed, and the day went along swimmingly, one finger and one eyebrow at a time..

The sky is the daily bread of the eyes.
Ralph Waldo Emerson 1803-1882

FUTURE 509. Most quiet need of the eyes:
Beagle Bay WA

Silence is a commodity that makes noise objectionable, but then it depends on the noise.

If you walk into an area of deep silence after having just left the sound of someone knocking seven bells out of a corrugated metal roof with a lump hammer, the silence will be the distinct opposite.

Out in the outback of Future Australia, on your home ground, the bashing of a corrugated metal roof will not be a noise as such; it will just be a pain in the neck end of communal disturbance, with kangaroos lifting or just flicking an ear, and dingoes pausing in mid-scratch for a moment.

It is weird. In the Sydney of noise and bustle of the oomph degree, the bashing of corrugated metal roofing is verboten, while the traffic on George Street creams along in an electrically driven procession of hushed common sense.

Future Australia will pass with flying colours into the future world of rational living

**FUTURE 510. Spitting into the wind:
Barwidgee WA**

If you call 'mekkin a pig's ear of things, as progress, we are doing great'. Words spoken by an old miner some years ago, a man who had nothing to show for his efforts, using the odd string of words to gather into a kind of esoteric 'why should we bloody bother' mode of social awareness.

Conversely, if you are doing an excellent job of things, then progress should be inevitable, and like the country of Future Australia, you will go far, and do well. Even so...

'You can guard against the idiot politicians some of the time, but for all of the time they will be there, and you will be spitting into the wind at certain times.' So said Claude Dampton as he picked himself up after falling over a spanking new wheelbarrow someone had left just outside his backdoor.

At first, he was a wee bit puzzled. 'Why the..?' Then he got it, and he smiled broadly as he thought of it. His friends had left him a 'barrowful of laughs' for his birthday. Ah, but... What *was* in the barrow?

What genus of contents of a wheelbarrow would make you smile, or even laugh?

If you are in Future Australia, the wheelbarrow on its own could make you laugh, why? How?

Questions that pose for answers but have no real meaning, so why should you bother, when to do so would mean you having to spit into the wind in a kind of literal sense. Can you sing, spit and duck your head at the same time? Now that takes a lot of skill and deft handling of verbal arrangements, and could well be beyond your verbal dexterity.

Eventually, Claude sat in the wheelbarrow and contemplated his slice of good fortune. He was a Future Australian, and he had time on his hands to consider the fact, and any other 'fact ' that might present itself in his happy sweet Future Australian life, with days that are longer and brighter. In Barwidgee, especially from within a clean new wheelbarrow - he laughed his socks off.

FUTURE 511. See the world from the deck:
Blair Athol QLD

I see the world from the deck of my electrically driven mobility scooter, which means I have to lower my sights a wee bit from my sitting position, and, for the most part, I like and enjoy what I see, and especially meeting with the good and kind folk, bless you all.

However, and I hated writing this next bit, there are those who are not only ignorant, they are also dangerous with their ignorance. I can perspire more common sense than they have, and thankfully, I have the ability to rise above such 'egoignomania' by at least three centimetres by lifting the right cheek of my backside, while thinking glorious thoughts of being in Future Australia. I can honestly tell you, the relief on such occasions is of a twofold magnificence.

FUTURE 512 Annie Faddi Flowers: (From Darwin) Andora NT

'There are a few more hours in life more agreeable than the hour dedicated to the ceremony known as afternoon tea'.
 Henry James 1843-1916

'Annie Faddi Flowers', a sign above a small shop in a backstreet in Darwin. It was a sign that gave the name 'Annie', the fact that she was a 'Faddi', and mad about he beautiful green and diverse colours and scents of 'Flowers'.

Annie was passionate about flowers, and she had always wanted her own flower shop, or Floral Display Shop as she preferred to call it, and her little place in Darwin had been made possible by her writing about flowers, and other decorative plants, of which she did not have one single favourite. To her mind, all flowers and plants were worthy of her praise.

There was one surreal aspect of Annie's little establishment, she served an excellent afternoon tea between noon and six in the evening, and it was the combination of tea and flowers that made the shop so popular with the folk of Darwin and around, and visitors from all parts of Future Australia.

Arranging a bowl of flowers can make a person feel relaxed and contented, and happy with a drink of tea, and a sandwich and/or a cake.

Annie, you are a gem, a Future Australian gem.

FUTURE 513. The children played to and fro:
Yandal WA

'It takes safe hands and time, for all to be fine and dandy'.

The grace and favour of life in Future Australia has made it fine for the young to be of good health, and having no hang ups of any racial or social kind, simply by them being Future Australians. Their young lives are 'longer' by them being kinder and understanding, and by the risk of the animosity of race or social standing being of no significance. *Just the fact that they are Future Australians makes them free of any social misgivings.*

Life in Future Australia is much fairer, and the children play to and fro in the peace of the land where they live, with the country itself in a better condition, and the people able to lead the world, a leadership that Future Australians take in their stride with ease.

Author's Extra Note:
Would that they would pass an act for folk to work and earn their bread,
Poverty would soon dwindle from the land and all ill feelings fled.

Would providence direct their thoughts to make such laws, and then, instead of outlawed folk, we might have free children, women and men.

Chapter 39. Green:

FUTURE 514. Under a rich sky:
Chorkerup WA

Under a rich sky, there is a rich soil taking care of the beautiful greenery, and Future Australia is blest with many variations of this greenery, all worthy of respect and care.

Take a look around, a proper look around, and do not be afraid to marvel at what you see in and around Chorkerup; a landscape rich in life, green and natural life, with the adornment of birds of all sizes, colours and songs.

Admittedly, at the time of 4.30 in the morning, the dawn chorus of birds singing might be a wee bit of a pain, and the wish for the birds to sleep in now and again would probably be the human 'dawn chorus'.

However, the full natural audience of Future Australia is something to be proud of, and getting up early in the morning 'now and again', won't hurt you, as long as it's no more than once a month say; you wish.

Even so, getting up early to get it right can't be all that much of a bad thing, surely.

Mother Nature and Mother Earth, our partners in Future Australia and Future World, have been getting it right for many thousands of years, and the dawn chorus of the birds has been part of that natural getting it right for the same length of time.

Therefore, you, my dear reader and fellow human, what right have we to mess things up. Blooming none, zilch, and that's me being polite in the future sense

Of course, we can be ultra polite and not 'swear', use correct grammar, and get blooming nowhere as the planet dies underneath our feet. I say bother to such 'pansying' about. This planet of ours was in trouble, deadly serious trouble, and for it to stay out of trouble, the whole world had to adopt the Future Australia way, the only way, or would you have rather pointed out the grammatical errors and say tutt, tutt? I blooming well hope not. You and I deserve better.

Chapter 40. Aborigine Style:

FUTURE 515. We all have a right. Allina's story Part 1: Baccabeyond QLD

We all have a right to be who we are, and to be as old as we can manage, there is no one on this Planet that is better than you, and we cannot say different, or Allina would soon put that right, and it has been thus for as long as she could remember.

Baccabeyond is a small hamlet at a bend in the road to Darwin, it doesn't matter where from, and to Delma it was and had been all she needed as a good place for her home to be.

Nestled in a tree-filled corner of the landscape, and far enough from the road to be isolated, Allina's cottage had fairly settled, hunkered down of you like, and Allina had done the same. However, just a few days ago, she had come across an old document that had been secreted away some seventy-five years before, when she had been just a wee child of five. It was a letter that indicated to Allina that her long ago ancestors were 'saintly', and that her mother's name, and her own, were taken from the old Raukkan Family of South Australia. However, the fact that she was far from saintly did not seem to matter to Allina, but she was most definitely Future Australian to the core.

FUTURE 516. Mortgage free:
Huskisson NSW

Some long time ago in Australia, let's say 10,522 years ago just to give it a number, an Aborigine fella stood by a small river in what was to become New South Wales.

He was looking for a suitable place to settle down with his new wife, the two of them looking forward to a good life together.

The woman allowed her home-making-instinct to decide where they were to live, and she chose a pleasant flat area that was above what they determined to be a flood level, and after taking a good look around, they decide together on the exact spot for their first home together.

On that very first day, they began to build their new dwelling, with one benefit they could not have had any clue about; they were going to be mortgage free, and they were going to build on free land.

Author's Extra Note: Although the ancient property has been long gone, the site is still there, it is still by the river, and the Future Australians who live on that site do so mortgage free, thanks to the Elders' adoption of the ancient Aborigine right of tenure.

FUTURE 517. Elders:
Caboolture QLD

Author's Extra Note: There is something very special about Australia - it's there. There is something very special about Australians - they are there too, and the real part of the new real world of the future.

Whether it be Old, New, or Future, it is the great country of Australia, and whether they are of Aborigine, European, or Asian descent, Future Australians are all great Australians. Of these it is the Elders and their system of guiding (Note: Not governing) that makes life in Future Australia a pinnacle of fantastic, and that's me being restrained by modesty, says he grinning from ear to ear.

Nevertheless, putting any sense of modesty to one side, if a country and its people are leading the rest of the world away from a potential disaster, from Armageddon in fact, then the term greatness is more than justified.

Question: What do you call a gathering of Elders?
Answer: A good idea.

FUTURE 518. How treasured their dwelling:
Merolia WA

To call it a house would be stretching credibility, but it was a home, a dwelling place built of local materials, some of which were once railway ephemera found and commandeered by Future Australian Dengo Smith, an Aborigine of Merolia who prided himself on skills passed down by his ancestors.

Of course, and it had to be 'of course', some of the materials made their way to Dengo's building site by the fortunate application of collecting by finding, it would have been left to rot or rust, or be blown away anyway.

It might have been a motley collection of bric-a-brac, but it had been transformed into a home, and Dengo was proud of the fact, and, to be honest, so was I, and proud of his ingenuity.

FUTURE 519. The years have given their gifts of kindness and time:
Cosmo Newbery WA

If we were to measure time by the passing of days and nights only, there would be a certain sense of gracefulness to life in general, and it is that kind of 'time-passing-life' the Aborigines have used for thousands of years, and now, in Future Australia, their lifestyles are integrated with other nationals' style of living.

As of Future Australia itself, in an oblique kind of way, many other nations have gained a great deal from their association with the Future Australian Aborigine non-political Elder system.

It is the Elders, of *all* nationalities, who have helped change a whole cultural mix into a profound singular easy-come-easy-go Future Australia, and along with this, is the vast improvement in the 'friendliness' and 'cooperation' of the different environments and nationalities found in Future Australia.

The years have given their kindness of time and made it possible for the natural world to be at one with the human population, safe and secure, with a plentiful supply of all that is good.

...

Author's Extra Note: Please try and keep in mind that this Planet we live on has been on a very fine edge, balancing between a global life of almost idyllic characteristics, and a cataclysmic planetary demise, and it is the humans who are holding the key, and keeping things on even keel, with the help of Mother Nature and Mother Earth.

Be scornful if you will, but you could help nudge things the wrong way, and quite easily, and there will be big lumps of low flying degradation hitting the fan, with nowhere to duck out of the way.

Think on.

FUTURE 520. Feel free to be happy. Allina's story Part 2: Badgingarra QLD

She couldn't give a gnat's doofer for school and schoolwork, and she could honestly say that much of it was just a waste of time anyway, especially on a warm sunny day in Badgingarra. On the sunny and warm days when she was less than 15 years of age and above 5 years of age, there had been the wood mill yard, where she and her friends found many secret corners where they could do things their parents did not want them to do.

It was fine and dandy to be young and free in those days, and Allinta relished out of school time when she and her friends could do what they wanted to do.

Those were the good old days of her youth, and like most of us, she had to grow up, be an adult, which she did, although somewhat reluctantly, but not before she had left her mark.

It was at the end of summer, and school was beckoning for her last year of statutory education, and she was far from happy about it.

No one ever made the connection, but Allinta knew who made a fire just by the school's wooden science lab, a fire that quickly spread through the whole school, a fire that made it possible for Delma and her friends to effectively leave school a few months early, and at that she was mightily pleased.

As far as she was concerned, if she said nowt, her education would not suffer, mainly because she didn't rate it anyway, and she did move on to be a good helper to her mother, on her good days that is.

FUTURE 521. The common touch. Sophia's life in Miriam Vale:

Miriam Vale QLD

The fact that she was of Russian descent did not mean anything to Sophia, in fact, because of her association with the local Aborigine children and their families, Russia, or any other country for that matter, played no part in her own mid-twenties everyday life in Future Australia.

After spending her teenage years helping the local Aborigine children to integrate and enjoy life as bona fide Future Australians, Sophia had found something within herself that was nothing short of wonderful.

In true Aborigine style, it took a long time for Sophia to be accepted into the social scheme of things Aborigine - the simple style of it was weirdly overpowering in a perverse sort of way. Then one morning she came upon two young ones, Tom (9) and Sally Berdinnon (10), who were of one of the local tribes.

Tom and Sally were obviously late for school; it was 10.30 in the morning. At Sophia's request, the two children agreed to accompany her to school, and they began to head that way, and they were nearly there when Sally spotted something.

It was a small bird feeding its young in a neatly constructed nest. Sally looked at Sophia and smiled, and in a voice that was so smooth and quietly sincere, she said; "That is *my* school, Miss Sophia, and we are not late if we are here."

With her eyes filling with moisture, Sophia could say nothing for a moment or two, and then, without any sense of dereliction of duty she said; 'Yes you are, you are here, we are all here'.

FUTURE 522. Commonness of common sense:
Leigh Creek SA

Putting the word 'common sense' on its head, we can perceive that the 'sense 'is the common denominator:

(1) The faculty or power by which external objects are perceived.

(2) The aesthetic response to an external object.

(3) A sympathetic response to an external object.

(4) Focus on the complexities of experience through perception.

To compare the previously mentioned with the acceptable designation of human common sense will bring about a definite...

That is where the part letter or part report ends. It was found in an old notebook in a shed near Leigh Creek in 1928, ten years after the end of the World War 1.

The only indication of the author are the initials 'A.F.G.'.

Author's Extra Note: I have included it in this work because of a tentative claim that 'A.F.G' was an Aborigine soldier/scholar. The source of this information must remain undisclosed.

I have considered the claim very carefully over a lengthy period, and my own tentative conclusion is that there is no reason why the author could not have been Aborigine, or as I prefer, Native Australian, some of whom I regard as good friends, even those I have yet to meet.

FUTURE 523. Tommy's peace and kindness: Buckleboo SA

In accordance with the previously mentioned, Tommy has a genuine gift; he knows how to give trust and peace to those in need of such valuable life enhancing features.

Future Australia is Tommy's country, and he is a Future Australian who us brim full of confident peace and kindness, though he will not suffer fools, and he cannot abide bad manners or disrespectful actions.

He is like a cool, clean and clear stream of humanity, sometimes grossly misunderstood, with not one malicious bone of thought in his body, and when an Aborigine family in a nearby village came upon really hard times, it was Tommy who gave up what he was about, and went to their aid. That was ten years ago, and now that family are Tommy's close and dear friends.

I don't suppose there is a lesson to be learned as such in Tommy's attachment to a struggling family, but when I learned of it, it did me good. It made me feel warm and comfortable in what I can only call a Future Australian way. In a similarly attached kind of way, I am also proud, if maybe a little stubbornly whacky, and thinking about that, I feel even more proud of Future Australia. and I would reckon there are lots of Tommy's in Future Australia to make me even more proud.

FUTURE 524. Twilight and the evening bell:
A Village in Future Western Australia.
There are chapel bells from a long while ago, and from a long way away, and some of these are now frail and tarnished with an age of inactivity. However, in Future Australia there are bells that somehow stand out because of their age, with a virile tone that sounds back through the years in a clarity that makes then seem close by, no matter where you are.
It is as if, or maybe it is an indication of how cities, towns and villages in Future Australia experience clearer and cleaner air, with sounds travelling nice and easy in the evening breeze over great distances.

The village folk have a smiling way of looking at the past. At exactly three minutes past nine every evening, a lone Aborigine rings the bell of an old chapel for eight ringing tones that drift in remembrance through the local and distant air. This is in remembrance of eight Aborigine men who died in a battle and war they knew nothing about, and for a long time their sacrifice went unnoticed, except by a few family members.

Then in the year 2034, a lone Aborigine was ringing the bell at the usual time, and as the bell sounded, folk from the village, and some who had come from a little further away, arrived to join in the remembrance.

Something had happened that signified a new social harmony; the new concord of Future Australia was being celebrated with the remembrance of those eight Aborigines who died for Australia, and in this harmony were the Future Australians who knew what it was to live in peace, and they needed to make it known that they cared.

Each hammer-to-the-bell sound went out to the memory of youthful Aborigines who knew not exactly why they had to die. However, they did know that it was expected of them, it was their duty, a duty they did not shirk. With their memorial being an old disused chapel in a new Future Australia, and in the year 2034, the bell was wrung by a young lone Aborigine who did know why he was doing so, and he knew he had to, it was his duty.

FUTURE 525. Remember:
A eucalyptus tree grows in the clearing in the outback, just a little way off the track. It has grown, on its own, right there for over one hundred and fifty-five years. It has developed from a

seed put there in 1916, and an Australian youth of Victoria who died that year in a far off land, was remembered by that tree, and he still is. Although he was not there when the seed was set, the youth is the fruit that made the tree.

...

Many folk have stood in Mortlake New Town, and many have marvelled at the ambience of the place, with a level of peace and tranquillity that is almost measurable by the attitude of the good townsfolk. They are not of a bombastic genre, they are more of a handshake and a grin kind of folk, folk you know you can trust, folk who are out and out Future Australian, and it does me good just to be here I can tell you.

FUTURE 526. Sunshine and rain:
Cowarie SA
The rain is of necessity when the ground needs to be of a condition to grow crops in Future Australia, the water thereby provided is of great importance, and is a modern miracle. After thousands of years of drought and sandy soils, the continent of Future Australia is now a massive park, and a land of plenty. If you are there, and you feel the cool soft rain on your skin, you will know what is meant be miracle.
Gradually, as the day opens its portals, the sun seems to smile down on the 'parkland', and in a friendlier way, no harsh searing desert heat, in fact, there are few desert areas, and when

there is a desert, it is so small as not to bother the landscape's new green fertile charisma of Future Australia.

Author's Extra Note: Life is a gift we should cherish, and in our living of it, we should always strive to survive.
The surviving of life should not be achieved by pressing down on those less fortunate; they may know more anyway, but we should do our best to lift them up by learning what they know. Think carefully before you decide, and even more carefully before you act.

FUTURE 527. Dreaming near the rocks:
Uluru NT
The huge monolithic Uluru is somehow a centre of power that a non-Aborigine will probably never understand.

All around it are the marks of secret footsteps unseen by the uninitiated, marks only murmured about by those who are of a tolerable ambition, and then possibly only in a dreaming state. It is the 'dreaming' that is perhaps a clue to Uluru's power, and if we accept the existence of this dreaming force of power, we can, with humility, be Future People of a Future Australia and of a Future World.

However, and I state this in all modesty, some could just think boo to it and not bother, and accept that the world will end, and very soon. Ah but, that is not the Future Australian way.

Yes, I know, the previously mentioned appears to be a big load of cods-wallop when you read it aloud, and clear. But it's the blooming truth, and it's high time we stopped messing about and become what we should be, out-and-out naturally good folk with a lifestyle in Future Australia to be proud of, and a lifestyle that goes all around this generous planet of ours. Future Australia is not a gimmick, it is a truth of this planet that we should fully comprehend and adhere to, and be aware, fully aware; the lives of our descendants depend on it, on us. The human race has got this far, to 2071, in a 'wrapping' of diligent cotton wool of tolerance from Mother Nature and Mother Earth, dither about now, mess up the opportunity, and we are really playing with a deadly firestorm of a potentially massive destructive power.

Future Australia is the only redeeming feature we have, do not waste it.

FUTURE 528. Under the quiet moon:
Canberra ACT

'A calm quiet heart is a continual feast'.

Future Australia is big, very big, and when the sun goes down it does so on the horizon that seems to be as wide as the world, with the night, and the moon that follows, seemingly also on the large side of big.

Out in the bush or the outback, the night dark is not quite complete when the sky, and the air above the land, are filled with stars, and of course other planets in a far away cosmos. Taking centre stage is the moon, which seems to be overseeing the whole scene with a soft glow of its ghostly white light spreading over the fast cooling continent.

Moving carefully, I settled myself to a slow walk along a track that was right friendly to me in the moonlight, smooth underfoot and well marked, a companion almost, and one I could trust fully as my feet moved from one step to another. Weirdly, I could sense the soft fall of booted feet behind me, but when I turned to say hello, there was just the light from the moon signifying my solitude in Future Australia, a solitude I could enjoy under the quiet moon, thanks to those who have gone before me.

Author's Extra Note: 'Get involved 'till the problem is solved', and then we can look eye to eye.

FUTURE 529. The compulsion of the dew: Queenstown TAS

It means nothing to a 'flat-footed goonybird' of the Tasmania Mountains, but when the dew forms on the ground at grass level, many animals breathe a sigh of relief, including the two-legged human species.

Dew is drops of water condensed on a cool surface, especially at night, from vapour in the air, and in Future Australia, it is this advent of dew on the grass that has served to make a huge difference to life. This is maybe not so apparent so much in the physical aspect of the dew, but there are ever expanding areas of good herbage for the dew to work its magic on.

It is like a secret society Mother Nature has contrived as a reward to Future Australians for their adherence to common sense, irrespective of whether the 'flat-footed goonybird' exists or not.

FUTURE 530. The lights begin to twinkle from the rocks: Meka WA

It is not to be expected that rocks have lights, or do they? Wait a minute though.

I was sitting on a rock near Meka in Western Future Australian, a rock 'bout the size of a small family car, when, in the gathering dark of the end of the evening and the beginning of night, right by my left knee, I spotted a light of sorts.

It was just a twinkle of a light at first glance, but bright enough to make me move my knee and take a closer look.

I suppose it could be the clean air of Future Australia that made it possible for me to see light emitting from a rock, and, I also suppose it could be that I am more susceptible, or of a more open mind to such phenomenon.

Ah, but, could it be that it wasn't a light at all. Could it be the action of moonlight reflecting off a titbit of silver or gold embedded in the rock?

Looking closely, and more carefully, the 'light' did begin to twinkle even more, and in a more certain fashion, and I became somewhat apprehensive. But then again, I could be under some kind of dreamtime enchantment.

Author's Extra Note: Maybe the peace-in-time has a way of making that twinkle of light a token of harmony, if so, I am more than happy to accept it in Future Australia.

FUTURE 531. Lying down in green pastures: Marble Bar WA

In Australia in the 20th century, the chance of a green pasture to lie down in depended on many things.

(a) Were you weary enough?

(b) Did the surroundings make you tired?

(c) Could you be bothered to carry on?

(d) Would you rather keep going until you found a better spot?

(e) Are you too thirsty?

(f) Are you excessively hungry?

(g) Are there any animal excretions about?

Then again, you could just lie down, think 'bother' to the lot of 'em, and go to sleep, in green pasture or not. But that said, in Future Australia, green pastures are a plenty, you are spoilt for choice, and you are free to lie down if you want to.

If it be that by green we mean lush short-cropped grass, your bed will be comfortable in the pleasant warmth of an evening.

But later on, when the dew begins to form in the cold of the night, please be sure you have sufficient 'bed linen', my dear reader. Maybe you will have some blankets, which will suffice until you can get to a drapery store - there's one just round buy the next bush-. Be aware, there is nothing more uncomfortable than wet bedding to lie on, and in. A blanket will keep the damp away, and you can sleep without the risk of contracting some fever or other if you are careful. However, you could adopt action (d) and walk all night, and then lie down in the morning absolutely knackervaraviched, or tired out, and oblivious to any of the aforementioned, in which case, good on yer.

Author's Extra Note: Words are like the wind, once read they can be blown away, and gone forever.

FUTURE 532. Lest we forget:
Pemberton WA

Author's Extra Note: 'A gentle heart is a continual feast.'
When the real truth of what they did hits you, let the sunshine bless you, and let the clean fresh air of Future Australia caress you, and may you know for real what they did, what they gave up for us.
We must never forget their sacrifice, and never forget our privilege and duty to remember them.
Pick up a stone, look at it, think of them, and lay the stone down again, gently.

Note: Just a few days before my 79th birthday,
I considered it a privilege for me to write the above,
And I make no excuse for feeling the way I do.
Bless them all, every one of them; and thanks to them,
My gentle heart is a continual feast.
May yours be also.

FUTURE 533. Fears may be untrue:
Ngukurr NT

Being afraid is not a crime, neither is it something to be ashamed of, and, if all things are taken into consideration, being afraid could well be the right way to go.

When I was considering, and then writing this futuristic masterpiece, (my definition) there was a sense of fear pervading in my mind, a question of huge consequential dimensions. What if my words fall on deaf ears, and the Planet continues on its path to total destruction?

You could be excused for thinking me to be a twit of the first order, but, and here is honesty for you, I don't really give a gnat's doofer, you think what you like, I know I am right.

On the other hand, if you are of the same mind as me, and you can see the truth of the fears, I look across this beautiful world to you, and wish you every success in conquering your fears with a gentle eagerness based on a sense of right.

Take heart from the fact that you, and I, by virtue of being afraid in a fearless way, we are saving, or helping to save,

Planet Earth, our home in the universe, and I say good on yer, good on both of us.

FUTURE 534. On that memorable scene:
Noonaman NT

If you see a neat pile of dog mess on the path in front of you, the sight of it will be ugh and sharp to the mind, then, seconds later it will be removed from your mind.

It's a crying shame that some memories fade, but then again, it is perhaps clever of us to have the ability to discount anything we wish not to remember, including the overwhelming evidence of the human degradation of our Planet for blatant selfish reasons.

The animals had no means to aid them to guard against the harm being done to them and the Planet by the human population's non-caring attitude; they are only animals, they will just die and become extinct, while the human population of the planet takes no heed, and lives on in ignorance.

However, Future Australia was and is different, and all life is considered sacred, possibly a lucky escape at the last moment for the whole Planet.

FUTURE 535. The fortress built by Mother Nature for herself:

Dalwallinu WA

We can prepare to engage in a fight for survival, but if that fight is against Mother Nature because of industrial/commercial or financial/commercial reasons, we will be perpetuating a sad bad fight that has been prevalent in the world for a very long and foolish time. However, with the onset of the Future Australian doctrine, the fortress built by Mother Nature has become strong, and Mother Nature can be resolute,* and the first real onslaught by her could well be a serious one and one we should do our utmost to avoid. Storms must be avoided if there is a threat of lives being lost on a huge scale.

It has become clear to those with common sense and true foresight, that the feeble humans have had it wrong in many ways, and some long years ago. But we do have a slim chance in a joint effort between humankind, Mother Nature and Mother Earth, a safe future for the planet will be assured, but it has been a very narrow escape route - we have been extremely lucky so far.

*Author's Extra Note: Mother Nature has just been practising so far, we would be unwise to test her too far. Unless we change our ways, there is a massive catastrophe to come, and we will be helpless.

FUTURE 536. The precious land set in blue oceans: Kelanie WA

I will get my lyrical pen out again, and mimic the old way of writing in the poetic way, with a nod in the direction of common sense maybe.

If we say Future Australia is a precious land, which I happen to think it is, then the oceanic waters around it could be regarded as the 'blue oceans', thus making true the title of this magnificent literary gem; my own defining inference.

Of course, I am more than happy to subscribe to such a slightly pompous affirmation of my talents as a writer, but, as always, the proof of the pudding is in the eating/reading, unless you are watching your weight; be careful with that one please. The 'pudding' in this instance is the continuing existence of life on this Planet.

FUTURE 537. The River ran and the Mountain stood: Kingoonya SA

Put your hand into a crocodile free river and let the water caress you skin, and experience one of Mother Nature's gifts.

Climb a mountain and when at the top take a look around at the view, that is another one of Mother Earth's gifts.

We are surrounded by, and live in a world full of gifts, little miracles often thought of as natural abilities, but make no mistake, they are gifts, and if we have them, then we should thank goodness we have. Furthermore, by using these gifts in conjunction with those that Mother Nature and Mother Earth are prepared to share with us, we can and should have a great life, all of us, and it is so blooming easy to do.

Future Australians are not in the business of dealing in nonsensical metaphors, or any other 'ridiculous' context with regard to social behaviour. Yes, of course, there are exceptions, but none are numerically bothersome enough to change the potential outcome of Future Australia and the rest of the world. Common sense folk far outnumber the pillocks in the Future Australian social scale, and because of this, what you are reading now is as near to the truth as I can get it, of that I am sure.

Would you care to be part of a successful Future social scene? If so, divest thyself of any pretence and self-advancement techniques based on 'stuff you mate', such as they are about as much good as a two-legged spider with cramp and no teeth. Stop messing about and get with it, be like the river and the mountain, and obey the rules of Mother Nature and Mother Earth.

FUTURE 538. If you can keep your head:
Sydney NSW

Sydney Edward Addison was a fit and able young man of 25, and he was proud of his work, what he worked on, and where he worked.

His natural ability was that of what some would refer to as a man-monkey, or 'monkey-man', his climbing ability was outstanding, and no obstacle or height of climb could stop him. If you can keep your head and climb with a sincere concentration, just as Sydney did, your chances of making it to the top and staying there are good to excellent. However, as in all ventures, especially of the physical kind, one slip is all it takes.

It was May 19, just four days before his 26th birthday celebration, and Joseph was at work, eighty metres up on the new Sydney Harbour Bridge, when, a fly, one solitary blooming fly, buzzed him, just in front of his left eyeball. A half a second of time passed, and by a trick of reflective action nd the breeze, he blinked, and the fly was trapped under his eyelid. The pain was excruciating and he...

Author's Note: There were sixteen men killed in the construction of Sydney Harbour Bridge.
I have the honour to list their names here.

Sydney Edward Addison 25
Nathaniel Swandella 22
J. Francis Chilvey 54
Henry Waters 50
Percy Poole 30
Henry Webb 23
James Campbell 40

William Woods 43
Robert Craig 63
Frederick Gillon
Alexander Faulkner 40
Robert Graham
Thomas McKeown 48
Alfred Edmunds
August Peterson 23
Edward Shirley

Not all died on the bridge, but it would be facile to distinguish one from another, and disrespectful.

FUTURE 539. The level sands stretch for just a little way: Geraldton WA

In the Australia of the 19th and 20th centuries, the outback land and deserts were large in area, and their propensity to evolve life was severely restricted by their huge barren expanses. However, in the Future Australian outback and deserts, the level sands stretch for a little way only, and in the search for a sustaining existence, the flora and fauna of these new smaller desert areas are able to live life to the full, and in an exuberant style, with the human element far outdone by the lesser creatures, but then...

'Having a bag of marbles for brains does have its drawbacks, especially when you have a brain fart'.

That less than delicate old Aussie man's observation of the human tendency to be finicky; if not downright blooming stupid in the implementation of the right and proper dedication of action, does in fact pinpoint the problem that had once plagued Australia and the world.

Please remember; and I reckon this to be bloody important, we were not at war with Mother Nature. She was at war with us a few times, and it was a close run thing, but thanks to Future Australians, we made it, but only just, next time we maybe not so lucky, so it is perhaps best we do not have a next time.

FUTURE 540. Of the forest's floor of flowers:
New Forest WA

The rich diversity of Future Australian flora and fauna in the spring is something else. I could say it is out of this world, but that would be a wee bit tacky, do you think, so I won't say it, eh.

Nevertheless, there are flowers of the Future Australia on the floor of the forests, flowers that do not exist anywhere else in the world, and their colourful proliferation is nothing short of magnificent, and, I can honestly tell you, being seated quietly amongst this magnificence puts things into perspective for me, a truly soothing and moving experience.

Dig out gold or silver and see it shine: a good description of frivolous falsehood, when all around there is more glamour than these supposed rudimentary ones can provide.

Not wishing to diminish the world of gold and silver, well, not too much I hope, I am within my rights as a self proclaimed Future Australian to ask the 'gold diggers' to look elsewhere for value. Of course, the 'digger' will probably tell me to go from whence I came, and that is his or her right, after all, you can't buy all what you need with flowers in a commercially driven society, but you can have a great life with them living and surviving on the same planet as you, and all the gold and silver in the world cannot buy that.

Author's Note: I found it difficult to write of an idyllic life in the future, but when I relaxed my early 21st century preoccupation with wealth and the getting of it, the task became easier.

Chapter 41. From a cottage window:

FUTURE 541. Looking at things in bloom in Future Australia from a cottage window:
Finniss Springs SA

It is a pleasant fertile valley, well wooded, and rich in green pasture, and my cottage at one end of the valley is my small haven of peace.

There is one door at the front of my home that is central to the facade there of. On entering, there is a passage from front door to back door, with a sitting room on each side, each room being four metres square.

Further to the back of the cottage are the facilities, with a kitchen and a stair to the upper floor where two small bedrooms will be found.

It is a small cottage indeed, but the building's outer and inner conditions are pleasing to my eye, and, as far as I am concerned, mine is the only eye that matters, as this gives credence to the gratifying of my inner being.

It is where I intend to stay; it is my home in Future Australia, a home from where I can spend time to look at Future Australia in full bloom.

FUTURE 542 Stand awhile in thoughts of peace: Gowrie Park Valley TAS

The touch of a thin and floppy leafy end of a tree branch can be surprisingly comforting on a still, hot and dry day. When Sophia experienced the lightness of touch from the branch of a huge eucalyptus tree on the side of her face, the surprise was at the state of ease and well-being that came over her.

Brushing the leafy limb of the tree to one side, she looked up and along the valley. It was so tranquil, and from her experience of what she could see, and looking at the hills around the valley, she realised she had never seen their equal, and there beneath the farthest hill, where it rises with such tree-filled grandeur, was the loveliest of cottages. It was the cottage-home of Leah Ningali, the Elder for the district of Gowrie, a simple but beautiful Future Australian dwelling that the old Aborigine she was about to meet had lived in for a very long time.

...

Author's Extra Note: If you say you are good at missing the obvious, what does that make you, honest?

FUTURE 543. In the middle of Future Australia: Simpson Desert

Trees, grass, and weeds, to the left of me, trees, grass and weeds to the right of me, and trees, grass and weeds in front and behind me and behind me. That is a long-winded way of saying

I am standing in the middle of a healthy landscape, on a level plain too large for me to estimate by a sight-and-see surveying technique. I am in what was once the Simpson Desert in the heart of Future Australia, in the year 2084, and it is a desert no more.

They may be just trees, grass and weeds, but they are green, and in the middle of 'Green Future Australia', as apposed to the sandy and stone nothingness conditions of around sixty years before.

Check this out:

(a) Tree:

Any large woody perennial plant with a distinct trunk giving rise to branches and leaves some distance from the ground.

(b) Grass:

Any monocotyledonous plant of the family Poaceae, having joined stems sheathed by long narrow leaves. Flowers in spikes, and seed like fruits; grass by any other name, or is there more grass than meets the eye? Yes, there is, and it means a great new future for Future Australia and the future world.

(c) Weed:

Any plant that grows wild and profusely without cultivation, and usually low growing.

(d) Nutrients:

Any of the mineral substances that are absorbed by the roots of plants for nourishment.

(e) Freshwater:

A clear colourless tasteless liquid that is essential for plant and animal life. In impure form it can be found in rain, streams, rivers and lakes.

A constant steady supply of clean water is all that we need to make the Australian continent green again. It's not beyond expectation, is it?

Author's Extra Note: Planting more trees and sowing more grass seed will pay a great dividend.

FUTURE 544. Dancing round the Maypole in November: Campbell Town TAS

Introduce yourself into the Future Australian Society, by that, I mean look 'em in the eye, and smile, it won't hurt you, or them. If you have come to the decision, and maybe the dedication of being a Future Australian, then, for goodness sake, be one.
You cannot afford to cherish pride or resentment and you should be of good cheer - which doesn't mean you need to imbibe in an over zealous way - and help others to get a good life by you setting an example by being nice and easy in everything you do.
Dancing round the Maypole when it's hissing down with rain is an idealistic attempt at normality in a difficult situation, so what, if that's what floats your boat. However, you could stay

nice and easy, and go and sit down somewhere out of the rain, and wait for the return of dry and warm conditions.

Whatever you do, be ready enough to do good for those you regard as being worthy of your help, and are worthy of becoming Future Australians, and if by any chance you want to be a Maypole dancer, good luck to you. It takes nerve to dance with bells around your ankles and knees, a handkerchief on your head and carrying a broom handle. I'm sure I couldn't.

FUTURE 545. It is a design of Mother Nature: Lawson NSW

Most towns in Future Australia have come about to afford living space for people, (not private cars), and 300 people in a small town may constitute a crowd, whereas 300 cars would not even fit.

The over 'population' of cars can make a township impossible to live in, in the normal sense that is. The only reason they do fit is the fact that folk have been brainwashed into seeing and using the car as a venerable object, and its security from damage or theft is paramount, therefore a safe parking spot must be found, and in the beginning of Future Australia they were parked anywhere the driver could find a space.

But then came the Elder solution:

It is a design of Mother Nature that a tree should reach up to the sky, and spread its branches, all in an orderly and even fashion. It was the decision of the Elders that cars should be parked in an orderly fashion in designated parking areas and spaces only. None compliance with the Elders' arrangement meant a tow-away of the car for scrapping, with no leeway for any excuses. This may be a brutal action, but it works for the good of all, and not just the motorists.

Author's Extra Note: Within a few months of the Elders' parking arrangements, the 'population' of cars went down by a third.

FUTURE 546. Jedda's story Part 1:
Cascade WA

Author's Extra Note: A settled mind is a worth many lives of a busy mind.

The rippling water of a stream without the presence of trees on the bank is just a mere run of water, good to see and hear. However, in the company of just one tree, with some of its branches over the water, the stream comes to life with such ease that it is a wonderful prospect to witness.

...

The water in the stream rippled along in flashes of silver reflections, and Jedda could sense every ripple, and while she was doing so, the tree she was standing by seemed to lean towards her in a friendly gesture from Mother Nature.

At that same moment, a shaft of sunlight blinked its way through the leafy branches of the tree overhead, and the rippling water became even more alive with flashes of light..

Keeping her back to the sun, Jedda placed a hand on the tree. Its strength and splendour enchanted her, and she smiled to herself with the joy of experiencing Mother Nature's handiwork so perfectly illustrated.

Over on the opposite bank of the stream, close to the water, a jamboree of small blue flowers seemed to dance lightly in the same soft breeze that was flicking at the curls of Jedda's hair, and as the flowers danced, their blue mixed in with the greens and browns of the grass and weeds. It would seem that the little flowers were part of a sweet harmony of friendly persuasion.

The effect was to make time to be of no importance as she watched the gentle display, and, still smiling, Jedda looked up at the tree and whispered a thank you to Mother Nature.

It did feel good to be in Future Australia on such a pleasant sunny day, a day she knew was going to be a good one, and giving a thank you did seem the right thing to do.

FUTURE 547. Take the lead, and go ahead:
On the Mitchell Highway NSW

You are not expected to agree with everything that is in my written work; that is your prerogative anyway. But at least, for goodness sake, get off your chair and do something for your descendants, children of the future that will, unless you and I do something genuine, (not the political bull), and try and create a world of fine sentiments and actions.

If you are a true Future Australian, I take my hat off to you, and good on yer. Take the lead, for goodness sake, take the lead, and go ahead.

You and your country are the FUTURE, the resources by which you can show the rest of the world how to live a good wholesome and honest life.

...

REMEMBER

It is well to remember that millions of good folk from Australia and all over this world died in conflict so the future of all could be safe and prosperous.

...

Author's Extra Note: I look over the land of Future Australia and see our chance, maybe our last chance, and I give thanks for that chance.

FUTURE 548. The litter buggers:
Kalbarri N.P. WA

Mother Nature is guilty of dropping litter, but it is all recyclable, in fact, her litter dropping is part of the magic of the natural world of the country, city, town, urban or outback.

Future Australia is chockfull of such 'country', and is thick with Mother Nature, with the only rubbish being that left by the uncaring dinkum-diddies, those who drop litter as if it is a requirement of showing how superior they are over other forms of life, something that was prevalent until the Elders took matters in hand in 2032.

That was the year when an Elder declaration made *all* of Future Australia an unnatural-litter free zone, and even Mother Nature seems to have complied, by her 'helpers' being able to give full attention to natural litter only.

Because of this intervention by the Elders, Future Australia is the epitome of a clean environment and social cohesion on being tidy naturally, for the sake of all.

Author's Extra Note: The habit of keeping tidy is an enjoyment that will last.

FUTURE 549. A blooming poem:
Cadoux WA

'You stand as much chance of writing a poem as I have of controlling a fart.' That, my dear reader, was said to me one wonderful evening, with a warm breeze rustling the pages of my notebook, the very same notebook I am writing in now, and my only means of solace was the hot mug of coffee sitting close by on an outdoor table.

To say my English companion was being sarcastic is perhaps making light of a truth I should have recognised some time ago. But being Future Australian, and looking further forward than your average English gonzo, I considered that my poem was worthy of recognition by a wider audience, after all, a poem is a poem, is a poem, is a poem..

However, on this auspicious day without my gonzo in attendance, and braving open criticism on all levels that would probably come my way, I stood up, notebook in hand and read out my poem.

My audience of four kangaroos did not even flick an ear, as my words drifted over and around them, and just as I made it to the last line, they gently turned and hopped away to fresh grazing, and probably better poetry elsewhere.

Author's Extra Note: The beauty of the evening; silent, with the glistening of the dewy air.
In a future land, with a blissful blessing, and me knowing that I was there.

FUTURE 550. I have never been in wealth:
Margaret Street, Sydney NSW

'Money can only give happiness, when there is nothing else to give it'.

Marianne Dashwood in Jane Austen's Sense and Sensibility. (Chapter 17)

Thinking on, as I am oft to do, the smile on my face must have widened as I penned the following:

'If only money wer' happiness, then I'd be a right miserable bugger.'

I have never associated wealth with happiness, simply because I have been wealthy once upon a time, and in a near millionaire way, and what I have seen of the more affluent folk and their possessions, I ain't all that impressed. I gave most of mine away. Therefore, after calculating cash, confidence and corruption, the 'non-wealth' I have now, makes me a better Future Australian and a better human being by far.

This is a very fine country, the valleys are comfortable and snug, the meadows are rich, and the deserts are either small or not deserts at all, and along with this, I am full of admiration for the beauty of Mother Nature in Future Australia.

Nevertheless, I am not all that keen on nettles, or thistles, or small creatures that bite or sting, but a troop of happy joyful village folk please me a lot more than the so-called wealthy, or stinking rich.

Author's Extra Note: If greed is the way to wealth, once you are wealthy, what happens to greed?

FUTURE 551. A good achievement:
Lake Violet WA

Disappointment can sometimes be like a swift kick in the nether region. However, my dear reader, the kick and its result can act like an aphrodisiac, so be careful of butt kicking, it can lead to exciting times.

Let me just declare, in the pristine world of cool thoughts and actions of Future Australia, life and liberty are mixed with truth and happiness, and on a level that some other folk find hard to comprehend. For these folk to arrive at the door of Future Australian common sense, and like what they see, it must be more uplifting than any other form of physical or mental alliance.

It bodes well for the world that the Future Australians have an inbuilt sense and desire to do what is right in anything they do, and the non-Aborigine folk have discovered and practiced this right way of live by learning from the Aborigines, even though it maybe in the eleventh hour. By doing so, they have changed the threat of global warming to the promise of a new and happier life for all, and on a naturally sweet blue green Planet, this is a good achievement, and one we can all take pride in.

...

My resolve is to ignore disappointments, and concentrate on achievements. For example, every morning I see the sun lift out of the far horizon of Future Australia, and my 79-year-old body begins to move. (Maybe more of a miracle than an accomplishment) It is as if is the first of such days; and there is more of the same to come.

Fifteen minutes before the hour of eleven a.m. and I am achieving a new recognition of life, walking, yes walking, to my favourite coffee bar, and not wondering if I will make it before the best bacon cobs in New Western Australia run out.
...
I made it, and a good bacon cob with a brilliant mug of coffee, my own pot mug, and I am sitting down and letting you know in writing how good life in Future Australia can be, and tomorrow I will be 80, good on yer Australia.

FUTURE 552. Unexpected delight and fulfilment: Harlin QLD

Now we are nearly there, sorry, now we *are* there, I feel I must embrace the Future Australian ambience of honesty, and a more forthright determination to achieve a better future for the Planet that is our home.

Where there was once the inability to join with others, there is now a strong nationally committed need and expectation to unite the different cultures into a natural togetherness with Mother Nature and Mother Earth.

The Native Australian way of life has been renewed, and embraced by all Future Australians, and the rest of the world, making the Planet safe and the future a better one. To give national delight in the promise of a good future for all who are Future Australians, and for all other peoples on the Planet who take heed and adopt the Future Australian way.

FUTURE 553. You are a world leaders; you *are* a world leaders:

Gascoyne Region WA

At last, the secret is out,

Native Australian.

Old Australian.

New Australian.

Future Australian.

If we assume the Native Australians to be the Aborigines, then the Old Australians would be the first soldiers, convicts, and settlers from Europe, mainly Britain.

Then came the New Australians, migrants from Europe and the rest of the world, to be followed with a mix of all three, to bring about Future Australia and the need for the rest of the Future World to comply with the Future Australians, the world leaders. If you are Future Australian, just take a look in the mirror and be proud to be a Future Australian.

Great balls of fire, being whichever makes no mind whatever, you are Future Australian, and whether you accept it or not, you and your fellow Future Australians *are* world leaders, and let nobody tell you different. On the other hand, if they do try to tell you different, tell 'em to 'go away' and go look in a mirror.

Author's Extra Note: Pull your socks up, and stride out as if you mean it, because, with you, the Planet stands a chance, a blooming good one, and it is the best and only way forward for all, so stick at it, please.

FUTURE 554. Are we there yet?
Arrabury QLD

'We are thereabouts there, but not quite in terms of actually being there.' That is a quote from an old document, a section of a written report made in Sydney in 1885. The statement was made at a certain meeting of civic leaders who were also church dignitaries. However, in my way of thinking, they were out and out politicians trying to use a thick pious disguise to hide a thin camouflage.

To squash any rumours to the contrary, if you are a Future Australian, you my dear reader are in a very good place, no doubt about it, and you can breathe a huge sigh of relief.

You are in a new country, a forward-looking country, where there is no finger pointing, or remarks from some snot goblin bureaucrats in an attempt to belittle and put a damper on your happy association with Future Australia. Just tell 'em to bollocks.

If you are black or white, or any other shade, you are fine. Blooming heck mate, you are in the same melting pot with everyone else, *every single other*, and you are so Australian it could make others less fortunate, not privy of your good fortune - welcome to Australia - and yes, we are fortunate, and we are there, every man, woman and child.

FUTURE 555. The wisdom to use the knowledge: Alambie NSW

Author's Extra Note: The ability to think and act with understanding and common sense, on the facts, feelings, and experiences, is to be known as part of humankind's intelligence. That is a long-winded way of suggesting the most favourable way forward, and by utilising this ability, Future Australian Elders have achieved a concise system of decision-making that makes the knowledge a global phenomenon available to all nations.

It was the Future Australian Elders who took the required steps towards a better world and a safer Planet, and it was the Future Australian Elders who nurtured the nation's people to be wiser and more understanding. They turned facts, feelings, and experiences into clear knowledge.

As far as I can determine, it is this re-formed knowledge that is keeping the Future Australia ahead of the rest of the world. Every Future Australian has the knowledge as if by a right of birth, they seem to know and understand instinctively the right thing to say, the right thing to do, and the right way to live, and the wisdom to use the knowledge.

Author's Extra Note: A blade of grass on its own is about as much good as a plastic frying pan, even if that frying pan is made of 'green' plastic.'

FUTURE 556. New streams and greener pastures: Aileron NT

The history of Australia was usually portrayed by hot sand and hot grit, with hot rocks and dry creeks, and thirsty conditions leading to the hotel door.

However, the Future Australia is a different country of new streams and greener pastures, with 'The Dry' being replaced with 'The Green', and seeing the sunrise through a canopy of soft green flickering leaves to signify this new clean manner of living.

Saying that Future Australia is an idyllic country is to be perfectly truthful, and given time, whole Planet Earth will be the same. A blue/green planet of perfect living conditions, the only one in the universe, and the one to take care of.

FUTURE 557. Remote Australian and the FBW: Canning Mills WA

A few distances south if Alice Spring, there is a remote pool of water that is there every day. However, this remote Future Australian pool of water is in a galvanised bucket that has long since lost its galvanisation, but this is by the by.

It may be an old rusty bucket, but it is always full of clean water, day and night, and year round, in that region of Future Australia. This 'FBW', Full Bucket of Water, is a lifesaver to many animals, and to some folk who wander by, creatures and folk that might be in need of liquid sustenance.

The FBW appeared one morning in September 2038, and since that morning, it has always been full of good clear water. There has never been an explanation for this phenomenon, but since that September morning, hundreds of 'FBWs' have appeared all over Future Australia, and their whereabouts is known to all who are in need of liquid sustenance.

How the buckets get there, or where the water comes from, no one knows, but they are now a permanent fixture in the green landscape of Future Australia.

FUTURE 558. Radical and Brave: Cadell SA

'Good change and bad change. No matter which, you need to be brave in the instigating'

That was in the old days, when being radical was being in favour of changes in political and economic conditions without reference to the social aspects of life in Australia, and these conditions were often applied against normal activities, and in the process swamping any chance of equalitarian progress. However, those days and thoughts are old hat, and they are no longer of any significance in Future Australian life.

The Future Australians are more pragmatic, and the notion of radicalisation is making inroads into the physique of all Future Australians, folk from all sides of the social aspects. Now, where politics and politicians played a huge roll, and took a hefty toll on everyday life in the form of stress and anxiety, there is a more placid and fairer system in place, with the Elders guiding the way forward with a lightness of touch that makes it easy for changes to take place in a kind easy-rolling fashion.

It is not so much a feeling of bravery but more of a sense of taking an easy part in a huge shift in Australian lifestyles.

FUTURE 559. The new one-way:
George Street Sydney

As Australians progressed from New to Future, one thing has stuck out as the progress progressed. Something just had to be done about the huge mess of bottled up traffic on the roads of cities, towns, and even in the outback areas of all places. Exactly how the thought permeated no one seems to know, but 'the thought' did permeate and turned into action, and this was the need to relegate the private motor car and the public bus to second and third range requirements respectively, to give people and the Planet a chance.

Stage 1:

Make all roads and streets one-way and make all roads pedestrian friendly as much as possible.

The idea was to take a leaf out of the 'Highway and Byway Rulebook of Mother Nature.' Animals in the wild do not collide head on; neither do they cause death by 'dangerous driving' as they move form 'A' to 'B'.

In the morning, they all move along their tracks in one direction only, and then in the evening, they all go the opposite direction. Yes, I know, that's maybe a slight over simplification, but it will serve.

Appendage to Stage 1:

Study a road and street map of any city or town, and see the lack of natural common sense used by the planners.

I could go on, but I think I will leave it there, and trust to human perception, as practised in Future Australia.

Author's Extra Note: Road deaths in Future Australia do not happen, not one, study that with a new perception if you will.

FUTURE 560. The Forest of Allendale:
Allandale SA

Author's Extra Note: As I walked through the Forest of Allandale, I could hear the flowers singing.

Make a connection with Mother Earth, walk by the trees in a forest and enjoy the natural energy therein, then sit quietly on a rock, or by a tree, and feel the essence of Mother Nature, and make your experience tranquil in a soul-satisfying way.

Refocus and allow your stress level to dwindle away to nought, and become a Future Australian as you make your minutes of relaxation expand into hours and days, then you can gain a fresh new Future Purpose in your life in Future Australia. Take a different grasp on life and really live.

Have the determination to carry through to a new and happier you, while at the same time the Planet gains in strength, with a new potential to give all creatures of now, and of the future, a safe home.

Take to the wild and feel the fresh breeze of Future Australia on your face, and know the same in the ground beneath your feet. Know the feeling, the sight, and the secret of a fresh new Planet, and be proud of your contribution. You are just as important as the world around you.

FUTURE 561. Tell it to the fairies:
Babinda QLD

'There is a fairy at the bottom of my garden called Nuff, Fairy Nuff.' That is an old phrase that transcribes myth and literary nuance without there being any harm done to human or fairy. However, it does leave a reservation. Is there some real awareness of the fairy population of Future Australia whenever the phrase 'tell it to the fairies' is used?

With a more natural ambience now installed in the Australian populace, it is within this writer's scope to ridicule and indicate the possibility of a fairy population in and around the human settlements of Future Australia, with the communities unaware. Dismiss the notion out of hand, deride any comment in favour, and scoff at any indication, no matter how obvious, and you could be risking missing out on the kind sensible influences readily available.

Author's Extra Note: In any event, please be careful how you live, and be careful how you perceive the way others live, and don't forget to smell the roses.

FUTURE 562. The beauty of the morning:
Yallingup WA

The most obvious pleasure of the morning is that you are there to enjoy it, to see the sun rise again on another Future

Australian day, and for you to witness the beauty of the morning in the guise of one who should be doing so.

It is your day, another day in your life in the fine country of Future Australia. Not only are you fortunate of where you live, you are also on a Planet that is of a new and more settled character, if a Planet can have a character that is, and in a truly universal sense of kindred spirits with all who know the Planet as a home.

Take comfort from what the Old and New Australians did for you by making it possible for a Future Australia to exist, and on a Planet you can rely on to be there for your descendants. It is now your privilege to be fair to the good folk of the future, and bestow a Super Future Australia and Planet to them.

Author's Extra Note: It's close, it's within our reach, don't weaken now, be strong, the Planet is relying on you.

FUTURE 563. The cool veranda.
Omeo NSW

'The man under the pergola of grapes'.

Osman, or 'Oz', is a man who hails from Turkey, the Turkey of some years ago, and he had made his home in Future Australia, in Omeo in the Snowy Mountains, from where he witnessed the change from New Australia to Future Australia, with some true sense of satisfaction.

He spent many nights in a double bed under a pergola festooned with grapes, with the over mantel of a seemingly never-ending Future Australian night sky.

Oz had spent a large period of his life wandering around the world, and it was at the beginning of his old years that he arrived in Australia, where he remained.

His Turkish upbringing was still with him, but in every other aspect he was Australian, Future Australian, and he loved it, especially when on his double bed on his veranda, watching the stars of a night sky from under the grapes on his pergola. Good on yer Oz, me old mate.

FUTURE 564. A plant has a life.
Huonville TAS

The fact that I can write, or put words in a particular order should I say, is down to the sacrifices made by folk in the Australian history of just over two hundred years or so.

Back n the 'beginning', many could not read or write. But dedicated individuals and groups, some of them a wee bit bigoted maybe, made it possible for young and old to learn to read and write, albeit on a slate with chalk. In time, in the gathering territories of the new country of New South Wales, eventually known as Australia, the balance between non-reading and writing, and the more learned sections of the new communities began to even out.

What the old bigoted folk did, should not concern us now. Any effort to uncover historical bigotry may be commendable, but it does no good for those who suffered under the yoke of the religious narrow-mindedness.

Out of prejudice comes common sense and the love of country and its folk, and this far outweighs any historical misdemeanours. Grab your rucksack, and warm up the old mobility scooter, and take a peek at the country that as been formed by the apparent mishmash of national fervour, through war and civil unrest, to a country of health and happiness, welcome to the country of Future Australia.

FUTURE 565. Mother Nature's ever changing course of Time
The Twelve Apostles VIC

Delicately, I place my ear and eye on Time. I do not hear it tick, I don't feel it move, but it does embrace around me in Mother Nature's ever-changing course.

Sunrise to sunset marks the passing of the slow and easy day, and sunset to sunrise marks the passing of the night, a gauge of Time that is equivocally one of Mother Nature's laws, and is of immense value to all of us on this Planet, and must be regarded with respect, and not just taken for granted.

Needless to say, Future Australia seems to have a huge day and night measure of Time. It may be the same 24 hour span used all over the world, but in Future Australia the distance from one horizon to another seems to spread Time out a wee bit, making the day longer and the night not so short.

It is not so much that you need a watch in Future Australia; you just need a good eye, and a clear and fair mind, and if you do happen to hear a tick, don't hang about listening for the tock.

FUTURE 566. The rags of the other end of Time: Ethel Creek WA

Unfortunately, even in Future Australia, to some folk the movement of Time is sometimes in spits and farts through the day, but because you are in Future Australia, such phenomena doesn't seem to matter. It matters not that the rags of the other side of Time are occasionally wafted in your face.

You may be standing by an old tree, one that had once reigned supreme over a broad expanse of nothing but tatty bush and huge patches of parched earth, whereas now, the tree is surrounded by lush pastures and the flicking of fresh green leaves of the many trees that grow in abundance thereabouts. If the previously mentioned describes where you are, take a deep breath of the rich oxygenated air about you. Enjoy the life-giving product of all the trees now growing there, and accept that Future Australia is like a huge reservoir of rich oxygen in the vastness of the Pacific and Indian Oceans, and is the leader in the virtual survival of all species.

Hot sun on its own is hot, but when filtered by the branches and leaves of vast areas of trees, the hot is changed to warm, and a pleasant natural warmth prevails.

FUTURE 567. Humiliated by generosity:
Rosemary NT

Abedel Guttrah had spent a lifetime, or thereabouts, walking, or should that be 'strolling' across a changing Australian landscape. His Aborigine upbringing had kitted him out for such a life, but he had reached his 'peace of mind' place. Roaming the village and looking out for a comfortable small cottage, and comfortable and sensible furniture if possible, he was taken aback. To his huge astonishment, he was presented with most of the furniture he needed as gifts. They were offers he readily accepted, even though he had never been so humiliated by such generosity.

All the folk in the small village of Rosemary went about their lives in a Future Australian temperament, in a village where life was seemingly composed in a symphony of genuine kindness and concern for others.

Abedel was truly humiliated by the kindness, and he was proud, extremely proud, of folk who had soon become his friends and neighbours, and he had smiled at his good fortune of being a Future Australian. To mark this state of local big-heartedness, Abedel planted as many trees as he could in a small dip in the land near to his cottage to show his appreciation.

After some years of real happiness, at an age heading towards one hundred years, there came the day, when, in a seemingly calm repose, he was lay down near his cottage, never to move again.

His life had spanned the move from New Australia to Future Australia, and to mark the fact the whole village made sure that the name of Abedel Guttrah would not be forgotten. They erected a small distinctive monument, and planted one hundred trees to compliment the trees he had put there himself, to form a pleasant wooded valley, and they converted his cottage into a museum of Future Australian village life.

FUTURE 568. Abedel Guttrah's legacy:
Rosemary NT

There is a small monument, a valley of trees, and a small cottage museum, three respectable assets that the pride of the small village of Rosemary.

However, Abedel's place of burial is a secret known only to a local Aborigine family and tribe, but his legacy is there for all to see and enjoy.

FUTURE 569. Molly's story:
Spring Hill

It was a story that Sandra's grandmother, Molly, often told her, the story of her first encounter with Australia.

It was April 1957, and the village of Swingate in Nottinghamshire in Middle England was going to lose one of its young women.

Molly Ormond was twenty-four and bored out of her lively brain, with a clerical occupation that was so mundane it would rank along with shovelling snow in a snow storm, she was getting nowhere.

Then came the chance to emigrate to Australia for fifteen pounds of British Sterling. This equated to about two weeks and half of her wages, and saving up took her just over a month, it was easy.

Ten months later, she was in Sydney Australia on her own, with an introduction letter to a women's employment agency, along with a permit to stay in a women's boarding house for two weeks.

She had two weeks to find a job and somewhere to live.

On the Wednesday of week one, she read a situation vacant column in the Sydney Herald. 'Help wanted to look after two children and do some housework.' The position was at a farmstead a hundred and fifty or so miles out of Sydney in the Parramatta district, and Molly answered the call as it were. and was soon being interviewed by a farmer and his young wife. Molly had landed a good position with a good Australian family, and she was eventually free to live a good life when she met her future husband. Their children grew up knowing only the existence of country folk in a flowering part of Australia, where the galahs swung and twisted in a cacophony of fun. Thanks to Molly, Future Australia has a family who had made it in a good way, a long way from where Molly begun her life.

FUTURE 570. The last word by Captain Cook:
Hicks Point NSW
Thursday 19th of April 1770
In words borrowed:
'It was squally weather and a large Southerly Sea. At six of the P.M. we saw land, and bore away along the shore Northeast of the Eastern most land.
Then we had the Southernmost in sight, at what I judged to be at 37 degrees and 0 minutes South and in the Longitude of 211 degrees 7 minutes West from the Meridian of Greenwich. I have named this Point Hicks, because Lieutenant Hicks was first who discovered this Land.'

FUTURE 571. Extra last word by the author:
Kimberley
Friday 2nd of August 2019 (249 years later)
In words of true concern for our sweet Planet:

We are now at a threshold where we take the common sense route, or we continue to follow the insistent politica-

commercialism of a so-called modern society, and ignore Mother Nature and Mother Earth.

We could dig our heels in, and say no to the enticement to commit to the destruction of our only home, our sweet blue Planet Earth, and know, really know, that we had at least done our best.

If you are reading my words some 249 years or so later, at least you know there was one who wanted to save you from the anguish of a dying planet. If I was correct in my assumption that Australia and Australians would lead the way to utopia, have a drink, and toast me in memory, along with those who gave their lives for us just over a hundred years before I wrote this .

Leslie H. Harvey of Kimberley.

Friday 2nd of May 2021

Printed in Great Britain
by Amazon